Energy Abundance

Introduction to Free Energy

First Print, April 2024

Energy Abundance | Karsten van Asdonk

Introduction to free energy

First Print: March 2024

ISBN: 978-94-6461-148-9
NUR: 961
info@obeliskbooks.com

Table of Content

"A refreshing and bold look at one of today's most pressing issues: energy. The book offers excellent solutions to the energy issue and points the way to a society that can carry these solutions. Indispensable input for the new generation of technologists and policy makers."

Dr. Bob de Wit

Professor Strategic Leadership,
Nyenrode Business University.

Preface

In March 2018, Karsten forwarded me his presentation on *free energy* that he gave in 2017 for Energy Now for the Studium Generale of Eindhoven University of Technology. He was 22 at the time and studying biomedical engineering.

I was pleasantly surprised. Speaking publicly about free energy at a technical university is not for everyone. In addition to knowledge of content, it requires a hefty dose of courage because the subject is the biggest taboo at technical universities. Nevertheless, he did a fantastic job, and I recommend that the reader watch the 9:11 minute video – coincidence does not exist – for himself before reading the rest of this book.

The reason for approaching me undoubtedly had to do with the demonstration of a free energy device that I had been privileged to organize at Delft University of Technology in 2010. This Yildiz magnet motor, also featured in this book, caused quite a stir in the ever-so-picturesque Delft, and the uproar reached the highest offices of the university. Once a video of the demonstration reached the internet, the genie of free energy in the Netherlands was out of the bottle.

Countless young technologists have since approached me for more information out of their drive to do something for the world. Many feel deep down that the current paradigms – especially those regarding our energy technology – need to be drastically changed. And they can. After all, none other than Nikola Tesla – arguably the greatest engineer and scientist in modern history – showed us the way. However, Tesla is hardly known at technical universities, if at all, even though we use his many inventions daily. It seems as if he has been swept under the rug.

What happened to me during that famous demo in 2010 was very telling. I was addressed by someone from the Delft Faculty of Applied Physics who asked me, "Mr. Vermeeren, this is interesting, but why are you organizing this?" Wary of the "official scientific paradigm," I explained that machines that evidently work should always be investigated, even if they perform feats that officially cannot be done. I received a striking reply: "We are working on this too, but we don't tell."

That statement stuck with me, but because the people above me strongly advised me not to organize such meetings anymore, it took six months before I sought him out. Then, in a frank conversation, he explained that TU professors indeed believe free energy exists but would never admit it to each other, let alone to the outside world. The reason is pure fear. Our systems, he said, are ideally suited to pressure and control people. This obviously does not benefit humanity and the entire world. When a choice must be made between obediently following along and enjoying scientific success or talking about essential issues and being ridiculed and fired, the choice is often easily made. It is a form of scientific hostage-taking. One thing became clear to me: we need young people who are not yet trapped between the narrow boundaries of the scientific establishment.

So, you can imagine that I was delighted with Karsten's email and presentation and that he decided to pursue a career in the search for the principles and workings of free energy.

This book is intended to free minds for free energy and thus help usher in the new paradigm. There is an incredible amount of information about free energy out there, but it takes a great deal of knowledge and persistence to separate the wheat from the chaff. All the more reason why, in addition to intuitive inventors (of which there are many), 'hard' scientists should also get involved.

Looking at the enormous potential at our universities and colleges, many thousands of talented young people must be ready to get started with this. However, they must be helped over a mental threshold. The subject of free energy must be taken out of the taboo. Readers can contribute by sharing this book with as many people as possible. The planet needs all of us.

Dr. Coen Vermeeren

Energy Abundance

Energy! A ubiquitous concept for all of us, but also one currently stirring up great controversy. While we take power from the wall socket for granted every day, like the air we breathe, the crisis surrounding our energy economy is coming to a head. We are starting to realize how fundamental energy is, because without it, everything would come to a standstill. In all respects, an abundance of energy is the foundation for our current way of life.

Energy is not only the basis of all human needs, such as clean water, food and a roof over our heads, but indeed all processes, products and services depend on a vast and constant inflow of energy. But where do we get this abundance of energy from, and what does it do to our environment and nature?

The short answer is that, generally, energy is generated in particularly destructive ways. Just think of the destruction of natural areas to extract fossil fuels like coal, oil and gas and the large-scale pollution associated with burning them. Hence, we are amid an energy transition: we need to eliminate the use of fossil fuels as soon as possible and switch to renewable sources – natural sources that do not run out.

In the Netherlands, the energy transition is currently characterized by two leading players in renewable energy: wind and solar. In 2021, 33 percent of the Netherlands' energy generation was renewable, three-quarters of which came from wind and solar.[1] Wind turbines and solar panels have the advantages that they hardly emit any pollution while generating energy and that they can be used in many places, but there are also numerous disadvantages.

The energy efficiency of solar panels and wind turbines is relatively low, so we need a lot of them. This means that fields must be filled with windmills and solar panels, for which large amounts of resources must be consumed. To increase the supply of renewable energy, it is now also the North Sea's turn to become a wind farm[2] with largely unknown ecological consequences.[3] On a cloudy, windless day, we immediately face an energy shortage, which must be supplemented with fossil fuels.

On the other hand, if there is too much sun or wind, there is a surplus, and the power grid becomes overloaded. Energy storage and "smart" distribution networks (so-called "smart grids") are proposed as solutions, but they, in turn, involve significant energy losses and complexity.

As we become increasingly aware of these drawbacks, calls are growing for a third alternative: nuclear power. The advantages: little pollution during power generation and a steady stream of energy. The disadvantages, however, are more than just trifles. Nuclear fission produces harmful radioactive waste, active for thousands of years, that we cannot (yet) process. Building a nuclear power plant is a massive operation that takes well over a decade. Nuclear power is not renewable; it still consumes a resource that is not infinite, be it uranium or thorium. History also tells us that nuclear power plants are vulnerable to severe accidents. Moreover, installing, maintaining and decommissioning nuclear power plants (in addition to wind turbines and solar panels) cost the Netherlands tens of billions of euros.

So, how will this energy transition proceed? By 2030, the Dutch government expects to get around 70 percent of its energy from wind and solar,[4] and there will be a ban on the use of coal for power generation.[5] By 2050, we should be completely off natural gas[6] while it remains to provide grid stability, and as of yet, half of our current power generation depends on it. The government is currently taking the 'first steps' for several new nuclear power plants, but they will certainly not be operational by 2030. They would only take over a few percent of total power generation per plant.[7] Meanwhile, large companies in the oil and gas industry are making record profits of billions of dollars a day due to high energy prices.[8] They have the greatest interest in maintaining this situation. The government is trying to tame the rise in energy prices with price caps, but the result is that burning gas is cheaper than heating our home with a brand-new sustainable heat pump.[9] All in all, we can say that the energy transition in this form is not working. But then what will?

The solution to our energy problem was found more than a century ago. Namely, a natural energy source exists that is everywhere and always present and never runs out: the ether. It is an energy field that resides in the background of the universe and extends throughout space. We can tap into this energy source to continuously make boundless energy usable anywhere in the world. This energy is free of pollution, free of limitation – and because it is always available everywhere – free of possession and free of profit. That is why we call it *free energy.*

Since the time of Nikola Tesla (1856 – 1943), countless scientists and inventors have succeeded in developing free energy technologies. However, these have never reached the market because they would drastically change the entire structure of society – especially the power structure. Indeed, an abundance of clean energy for the average person means the collapse of trillion-dollar corporations[10] in centralized energy. As these "world leaders" push for more scarcity and austerity – turn your thermostat a little lower to save – while we keep paying more and more, humanity's true potential lies in unprecedented abundance and freedom.

Why don't we hear about this? There is quite a lot at stake. It is the total shift of power, the change of all our systems and our beliefs of what is possible that make free energy so controversial. As forces of big money, energy giants will have none of it because they would lose their right to exist. Governments don't want to share anything about it, supposedly in the "interest of the citizen," because it would threaten the economy and national security. Established science does not want to hear about it because it would have to admit that some fundamental assumptions about the laws of nature must be corrected, and the books would have to be rewritten. In short, free energy faces the greatest resistance at all levels of society. The good news is that we *are* society – we have created, are part of, and authorize all the above institutions. It is, therefore, up to us to open up to free energy.

This automatically points us to the inner dimension of the free energy phenomenon. Whether we are ready for such changes can only be answered by turning inward. Who and what are we really, and why are we here? What kind of world do we live in, and in what kind of world do we want to live? Throughout this book, it will become clear that this inner awareness is crucial to developing technology such as free energy. Indeed, abundance is not just a technical feat. It is the true nature of the entire universe and, thus, inherent in who we are.

This book – an introduction to free energy – offers a perspective on this world of abundance. We will discover how free energy can work, why it has not yet broken through, and its grand implications on every aspect of our existence. We will also shed light on what is needed now to manifest free energy and how we can contribute to it.

The book is written for all scientists, entrepreneurs, policymakers, students, artists and free thinkers, professional or amateur, who want to commit themselves to a more beautiful world. Hopefully, it will motivate you to get started with free energy yourself, whether you are still in school or already enjoying retirement. However, the book is also here for those who, out of sheer curiosity, are in for a fresh and positive perspective on society, technology and the future. Inspired by the information in this book, we cannot help but be optimistic and confident that we can inhabit an unprecedentedly beautiful world.

1 www.cbs.nl/nl-nl/nieuws/2022/10/meer-elektriciteit-uit-hernieuwbare-bronnen-minder-uit-fossiele-bronnen

2 www.theguardian.com/environment/2023/apr/24/european-countries-pledge-huge-expansion-of-north-sea-wind-farms

3 www.nature.com/articles/s44183-022-00003-5

4 www.government.nl/latest/news/2020/12/04/north-sea-energy-outlook-establishes-framework-conditions-for-future-growth-of-offshore-wind-energy

5 ieefa.org/resources/ieefa-update-netherlands-new-program-close-all-coal-fired-generation-2030-sends-european

6 www.oxfordenergy.org/publications/the-great-dutch-gas-transition/

7 www.world-nuclear.org/information-library/country-profiles/countries-g-n/netherlands.aspx

8 www.theguardian.com/environment/2023/feb/09/profits-energy-fossil-fuel-resurgence-climate-crisis-shell-exxon-bp-chevron-totalenergies

9 nos.nl/artikel/2445463-consumentenclubs-bang-voor-vergeten-groepen-bij-prijsplafond-energie

10 www.statista.com/statistics/664868/global-top-energy-companies-by-revenue/

11 zetookdeknopom.nl/

Chapter 1: Introduction

An abundance of clean energy is hard for many people to imagine. This is unsurprising in today's zeitgeist of perceived scarcity and crisis. Therefore, the concept of free energy raises many legitimate questions that we will discuss throughout this book. Let's answer the most frequently asked questions in advance below and highlight why free energy is so important right now.

What is free energy?

Free energy is any form of energy – like heat, motion or electricity – that can be generated endlessly and cleanly at any time or place in the universe. This makes free energy *the* answer to what we call a global energy crisis today. The technology that makes free energy generation possible is called free energy technology. It requires little to no polluting fuels, produces hardly any waste and does not rely on non-constant energy sources such as the wind or the sun. Instead, it taps into an energy field that is always present everywhere.

Imagine a device the size of several shoe boxes. Instead of a plug to plug into an outlet, it has its own outlets – continuously generating its own electricity. With this device, you can charge your laptop and phone indefinitely, run your washing machine and keep all your lights on. Since such free energy technology is scalable from small to very large, we can provide unlimited clean energy to every home, car and city, independent of a centralized energy supplier. The term free energy thus implies a new level of *freedom* in managing our energy economy: a continuous source of clean energy is available anytime, anywhere.

Does free energy exist?

Devices that can supply unlimited energy may sound too good to be true. So, why aren't they here yet? For most people, simply "seeing is believing" applies. However, as we will see, people who witness the working of a free energy device with their own eyes often don't believe their own eyes. This, too, is unsurprising since most have never heard of it, and it goes against everything most people have learned.

Therefore, this book is not primarily intended to convince you of the existence of free energy. It does serve to share knowledge and passion about it with you. Wait to take everything discussed for true or false immediately, but be inspired to do further research to draw your own conclusions.

How does free energy work?

For thousands of years, the notion has existed that everything in the universe has to do with energy. Everything we can see, and touch is, in fact, a form of energy, as are all the forms of energy we cannot see. Consider, for example, the radiation from your Wi-Fi or the heat radiation from a heater.

Renewed insights into physics show that even a vacuum – the empty space without atoms or molecules – is not empty but is filled with vast amounts of invisible energy. This so-called vacuum energy, which we find everywhere in the cosmos, is also called the *ether*. The ancient concept of the ether, which was in vogue until the end of the nineteenth century, has been temporarily cast aside in modern physics. Yet the ether seems responsible for sustaining all other visible and invisible forms of energy, from all the atoms in your body to the largest star systems in the universe.

This ether is not a rigid substance but continuously in swirling motion. It is the energy of this movement that we can convert into electricity, for example, using special equipment. You can compare it to an invisible wind that can be converted into electricity by a windmill. This "ether windmill" needs no fossil fuels, and because ether is present everywhere in the universe, this energy can be harvested anywhere.

We do not know where this ether comes from any more than what came before the Big Bang. However, the universe is so gigantic that we can safely assume that at least ether energy does not run out. Moreover, we cannot yet answer whether ether energy is 'consumed' in the first place. Perhaps energy cannot 'run out' altogether in that respect. The universe seems to operate in infinite abundance and certainly not in scarcity.

This is only a brief description because there has yet to be a consensus on precisely what the ether looks like and how ether technology works. Chapter 3 delves deeper into physics and technology to better understand how free energy might work.

Why isn't free energy technology here yet?

If free energy is so evolutionary and groundbreaking, why are we seeing and hearing next to nothing about these technologies? The answer is that the world of free energy is shrouded in mystery and controversy precisely because it is groundbreaking. With the introduction of free energy, everything is going to change. Not just the energy and transport sectors – by far the most essential and energy-intensive industries in the world – but also everything else that depends on energy. There are so many interests involved, not least financial interests and stakeholders, that it is rightfully called *disruptive technology*. It has the potential to overturn the entire established order.

Anyone who studies free energy carefully and reads the history of pioneers who have been working on it for more than a century can conclude that numerous inventors have succeeded in developing free energy technology. Unfortunately, however, these people were far ahead of their time, so most of their conservative contemporaries did not appreciate their work. Jealousy, ridicule and exclusion were no exceptions. Added to this were highly influential stakeholders who benefited from the status quo. We will see examples of these factors later.

To this day, humanity apparently has not been ready for free energy. However, more and more free energy research is being openly discussed today. The internet proves to be invaluable in this regard, both in finding information and in sharing it. But as with everything, sense and nonsense coexist in the field of free energy, so it takes sound knowledge to separate the wheat from the chaff. Through the experiences of inventors and highlighting the large-scale changes that free energy will cause, it will gradually become clear what kind of resistance this technology faces.

The importance of free energy

If the mainstream media and current political discourse are to be believed, humanity is headed for a massive climate disaster unless we take action now. That action will cost many thousands of billions of dollars and euros, but not only money. Some of the measures are also at the expense of the very nature they are supposed to protect. Look, for example, at the large-scale wind farms that threaten the environment and animals or at solar parks where nothing can grow or flourish except the panels themselves, not to mention the burning of primary forests as biomass. But are the premises of the so-called climate threat even correct?

The climate is changing; that is a given. After all, change is the only constant in the universe. Exactly how humanity influences Earth's climate is a story on its own, and contrary to what we are told, in the scientific world, there is no consensus on that at all.

What is true is that humanity, with its current technological development, has gone down a blind alley. Man is polluting and destroying nature in unimaginable ways; he depletes her raw materials, cuts down her primeval forests, overfishes her oceans, pollutes air and water and plunders her soil. The bulk of the production and consumption of goods is set up linearly: it moves irreversibly from A to B, from cradle to grave. Humanity lives in imbalance with the nature of the Earth, and this imbalance must be restored. Everything we do should cooperate with nature and not go against it.

Never before have we been so pressed by the fact that energy costs lots of money. The energy crisis that started in 2022 will be a game-changer in our global society. The 'solution' currently proposed is based on the idea that energy is scarce. The adage is that energy generation is a "problem," that energy is "expensive," and by definition, "polluting." The common man who uses energy is thus made to feel guilty. Instead of energy no longer being a point of discussion – because the energy issue has been solved – energy has taken center stage in social discourse.

Today's energy technology – generating electricity, especially for transportation, heat and production processes – is unimaginably polluting. However, we cannot do without it. In addition, energy is also a very profitable business model for companies. New competitors in the energy market are closely watched by stakeholders, and changes are, by definition, met with resistance. Genuine solutions that serve society but disrupt financial and geopolitical forces are nipped in the bud. We have unfortunately seen many examples of this, some of which we will discuss.

Regarding pollution, fortunately, more and more recycling has been possible in recent decades. However, this is still not truly viable because it costs a lot of energy in most cases. These costs are rarely passed on when a product is purchased because it is less competitive against producers who do not take pollution so seriously. Buying a new product is often cheaper than repairing or recycling the old one. Renewable alternatives often remain more expensive than fossil fuels in generating our energy, so they can only ever partially replace the old systems.

The idea of free energy very quickly shows that this vicious cycle can be broken once and for all. Once clean energy is available in abundance anywhere and anytime, energy is no longer an issue. It becomes easy to produce and recycle cleanly. Other far-reaching implications include being able to desalinate seawater for an abundance of clean drinking water. There will also be abundant energy to repair our damage to nature. For example, it will become feasible to clean up all the plastic in the oceans – something very costly today, mainly because of the cost of energy. Meanwhile, we can travel and transport our goods with minimal pollution. These are just several ecological implications. In Chapter 6, we will further explore the endless ramifications of free energy.

The spiritual dimension

In addition to a physical dimension, free energy has a deeper, spiritual dimension. The word "free" is therefore well chosen. Free energy gets to the heart of living on Earth as a free human being. Free energy is free of pollution, free of interruption, virtually free of charge, and independent of third parties (giant energy corporations), geopolitical relationships and resource wars.

Although some of us might say we live in great freedom relative to other parts of the world, we are most likely utterly dependent on an enormous, centralized power structure. Energy is a crucial factor here.

As long as energy remains available only through a centralized grid, we are forced to participate in destructive ways of energy generation and must join society's treadmill to pay our energy bills and "survive." Of course, we always have a choice, but if we decide not to participate, our lives will not become any easier.

Free energy allows citizens and communities to regain control of their energy management. The top-down structure of energy generation can be broken down, and a new bottom-up structure can emerge, in which we can freely and without blame or shame decide how much energy we put into what. Energy will become a fundamental right, available to everyone everywhere and always. We will no longer have to fight for it. As we will see later in this book, this significantly impacts our lives. We no longer must fight to *survive* but can indeed start *to thrive*.

From survival to thriving

So, what does it mean to thrive? That is a question everyone should ask themselves, but first, it is helpful to see where we are still in "survival mode." Our survival instinct has naturally brought us to where we are today. However, that survival instinct has also created countless (useless) wars and conflicts. These are extreme forms of survival, but they also play out in subtler ways in our daily lives. In business, for example, we must constantly compete and become better than the rest to be successful. A healthy degree of competition is stimulating, but it overshoots the mark when it comes at the expense of connecting with each other and nature. Before we even realize it, we are in direct competition with nature herself and must force her to obey – otherwise, we might die!

From early childhood, we are instilled that this is how the world works, but that is only a matter of perspective, a story about the world. The story only becomes a reality when we collectively believe in it.

Let's do an exercise. Relax and take a few deeper breaths. Imagine a society that has figured out the game of necessities completely. Everyone has, cheaply or for free, shelter, clean energy, clean water and healthy food. All these processes are realized sustainably by people who have made it their passion and are in harmony with nature. In what way would you like to fulfill your life now? What would be your calling?

This little thought experiment is a glimpse of the reality accompanying the introduction of free energy. Since energy is the basis of our existence and survival, abundant clean energy will ensure that we no longer need to worry about whether we will continue to live. We can then deal with the more significant questions of life, such as: Who am I? And why am I here?

Remarkably, this chain of cause and effect also works the other way around – or perhaps even better. Once we are no longer concerned with survival and ask ourselves these more profound questions, our consciousness expands, and we become more in tune with the abundance inherent in our universe – the perfect fertile soil for a technology that reflects this abundance. As we ground ourselves deeper into the realization of consciousness, we will begin to look at our world differently. Therefore, our "level" of consciousness goes hand in hand with introducing free energy.

Paradigm shift

"There are more things in Heaven and Earth, Horatio,
than dreamt of in your philosophy."

William Shakespeare

As you can see, the integration of free energy means a transformation to a new human civilization, with a paradigm shift at every level. Let's dwell on that for a moment.

Sooner rather than later, every aspect of what we now accept as "normal" will be broken down, transformed or renewed. In principle, a paradigm shift implies a revolutionary scientific breakthrough. Science methodically tries to figure out how the world works and what reality is, and over the past few centuries has increasingly determined our worldview. For example, when a motorized airplane took off for the first time in 1903, the prevailing paradigm that a vehicle heavier than air could not possibly fly shifted, and it was proven against all odds that a new era of travel, transportation – and unfortunately, warfare – had arrived. A paradigm shift is a sudden leap in the development of our perception of the world.

In this way, the introduction of free energy will represent one of the greatest paradigm shifts in human history. In the process, science will reunite with spirituality – a connection severed in the Age of Enlightenment when religion and science were split into the institutions of church and university. By spirituality, we mean the notion that the entire universe stems from the same source of consciousness and that everything is somehow interconnected, an unbroken whole. This notion is underpinned by the advent of quantum physics, the implications of which have yet to dawn on humanity.

The moment free energy makes its full entry into society is the moment in history when our economy, ecology, politics, religion, science and all other facets of life will change. We will then enter an era of renewed connection between ourselves and nature and even between us and the entire cosmos. This may seem utopian, but step by step, we are getting closer.

In this book's final chapters, we will explore this vista again. Let's start with one of the greatest pioneers and the primary source of inspiration for writing this book: Nikola Tesla, the "founder" of free energy.

Chapter 2: Nikola Tesla – Founder of the free energy movement

Few people know who he was, but many of us deal with his work daily: Nikola Tesla. We owe much of our current electrical technology to him, and his work enabled humanity to make a giant technological leap at the end of the nineteenth century. Tesla was a Serbian engineer with a particularly visionary mind. He not only provided the foundation for alternating current, radio communication, X-ray photography, radar and the induction motor, but he is also considered the founder of the free energy movement. His greatest desire was free energy for the whole world, and he devoted much of his life to making it an experienced reality for everyone. He failed. Was he perhaps too far ahead of his time?

Although Tesla's work became the foundation of our society, he ultimately did not receive the credit he was due. Even in the current scientific community, he is hardly known. Yet besides being a brilliant engineer and visionary, he was a celebrity in his own time – a man who captured the imagination of many despite his rather reclusive life. You will find his vision of the future in quotes throughout this book. What was that vision, where did he get his inspiration from and what has become of his most important inventions? That is what we will explore in this chapter.

Who was Nikola Tesla?

"The desire that guides me in all I do is the desire to harness the forces of nature to the service of mankind."

Nikola Tesla

Nikola Tesla (1856-1943) was born in the village of Smiljan, in what is currently Croatia. Inspired by his mother, who descended from a long line of inventors, he was very curious about technology in his youth. His father, who had a career in the military and then became a priest in the Serbian Orthodox Church, was a strict man. He tried to steer Tesla in the religious direction early on. The family lived a simple village life on a farm with ample land, where Tesla could let his imagination run free as a child. This changed abruptly when he witnessed the death of his older brother, who was trampled to death in the barn by one of the horses, which was frightened by a loud thunderstorm. Tesla regularly dreamed about his brother's bones breaking, which sounded as loud as the thunderstorm itself. The playful young Tesla became introverted.

Figure 2.1: Nikola Tesla circa 1890.

There was a second reason for this. Tesla regularly had very vivid hallucinations accompanied by violent, uncontrollable flashes of light in his mind's eye. But besides initially disturbing him, this caused him to develop a rare skill. He only needed to hear a description of an object to see it vividly before him, only to find out it wasn't there when he ran his hand through it. Later in his teens, he called this condition of body and mind "a gift from God." He had trained himself to direct these "hallucinations" by sheer will and mind power.

He became a master of his gift and was eventually able to visualize whatever he wanted to see in razor-sharp detail. Visualizing new worlds and meeting new people in cities he had never heard of became his hobby. This would become his greatest tool in creating his inventions, for which he never had to make drawings or models beforehand. When he was working on an invention, Tesla could build it in his mind, look at it from all sides, test it and improve it as needed – all in his crystal-clear imagination. In real life, his inventions were precisely as he had imagined them.

After studying engineering and physics, he worked in Europe's then-emerging telephony and power station industry. Tesla then moved to the United States in 1884, according to Tesla, "the land of the golden promise."

In the United States, Tesla would go on to work for his great idol, Thomas Edison. At that time, Edison had already made a big name for himself with his research into electricity and his electronic inventions. However, he used direct current, a technology with significant practical limitations that made his equipment less functional. Tesla came up with a solution: alternating current. Edison was skeptical but wanted to give the young European engineer a chance and offered him $50,000 – a sum of $1.5 million today – if he could improve each of Edison's electrical devices. Tesla succeeded in doing this in record time, but Edison didn't live up to his end of the bargain. He called Tesla a fool for having taken his offer seriously and paid nothing more than his weekly wage. The disillusioned Tesla immediately left Edison and subsequently found himself financially grounded. He was even forced to dig trenches for two dollars a day to make ends meet.

However, his genius did not go unnoticed. With the help of two businessmen, he soon founded the Tesla Electric Company. In his company laboratory, he invented the electric induction motor that can still be found in countless devices today. This motor – which could also be used as a generator – laid the foundation for efficiently generating alternating current.

He later sold his alternating current and motor patents to George Westinghouse, who aptly recognized that Tesla's invention would change the electric industry forever. In this way, Westinghouse became a crucial business partner and friend of Tesla, enabling him to set up his own independent laboratory. Finally, he moved to New York to throw himself into his latest ideas, which would surpass all his work up to that point.

Radiant Energy:
A new form of electricity

Impulse electricity

While scientists and engineers were still busy introducing new alternating current generators and motors, Tesla accidentally discovered a unique electrical effect when he experimented with the discharge of high-voltage direct current. He found that as soon as he shorted a charged high-voltage capacitor with a switch, the abrupt discharge generated an *impulse*. The discharge was comparable to a small lightning strike and was accompanied by a loud bang.

The effect of such an impulse was felt as a powerful shock wave, which penetrated both insulating and conductive materials. Even though Tesla stood behind shields of glass or copper, he still felt a stinging pressure from the electrical discharge – as if he had been shot with tiny needles. That was bizarre because grounded copper blocks most forms of electromagnetic radiation, and a glass shield should have ensured that any pressure wave from moving air would be stopped. Instead, the discharge apparently moved particles right through all kinds of materials. That would mean it had to be a field effect. A hitherto unknown field was set in motion – of which Tesla suspected it must be a form of ether – and passed straight through all matter and had an influence even at great distances. Tesla called these shock waves generated by impulse electricity *Radiant Energy*. In the next chapter, we will look closer at the energy field, the ether, that Tesla set in motion with his Radiant Energy.

Tesla developed ways to generate Radiant Energy in a controlled manner and to focus it without the adverse side effects of a stinging shock wave. Even at a distance, Radiant Energy could electrically charge conductive surfaces. Thus, devices with a suitable antenna could be supplied with this energy wirelessly and efficiently. Even more significant, however, was an effect by which Radiant Energy amplified the electrical power in a circuit without Tesla having to add energy himself. We will describe a few of Tesla's inventions below to understand better how he worked with Radiant Energy.

The Tesla coil

The famous Tesla coil is essentially a transformer with two coils that transforms a low-voltage direct current into a very high-voltage alternating current. The second coil has the most windings and thinner wire and generates a strong alternating electric field. Using the strength of this field invisible to the naked eye, Tesla regularly theatrically demonstrated how his coil produced enormous lightning bolts and lighted lamps without having them connected to a power source with a live wire. Although this was a spectacular sight in the public's eyes, it wasn't what Tesla was *really* excited about. This is because the strength of the alternating electric field decreases rapidly with distance from the coil. Even a hand between the coil and the lamp could nullify the effect of lighting it wirelessly. The lighting of wireless lamps and the spectacular sparks were, therefore, just for show. Tesla was primarily concerned with Radiant Energy – energy that could be efficiently transmitted over long distances. Over time, confusion arose about what the Tesla coil was supposed to do. As a result, significant differences have emerged between Tesla's original design and the Tesla coil often being replicated today.

When Tesla was not giving demonstrations to the public or investors, he was mainly set on minimizing the massive shower of sparks that the coil could generate. Indeed, much energy was lost in producing those sparks and the electric field that could light lamps. Spectacular, but not very efficient.

Tesla experimented with different types of coils. In addition to the cylindrical coils shown in the images above, Tesla used cone-shaped, flat, and bifilar (double-wound) coils, among others. Tesla invented and patented the latter type, one of more than 500 unique patents to his credit. At this time, we are still unsure which kind of coil worked best.

Tesla also used thousands of sharp impulses per second to make the coils resonate – meaning, to vibrate electrically at their proper frequency. He noted that the impulses' voltage, frequency and "sharpness" needed to be very high for the Radiant Energy effect to emerge. This is usually forgotten nowadays and is technically challenging to reproduce.

Figure 2.2: Left: a modern variant of the Tesla coil, which unfortunately does not generate energy or transmit it over long distances but does produce large sparks (Photo: OneTesla). Right: a variant of the Tesla coil built by the author. The fluorescent lamp is lit without being connected.

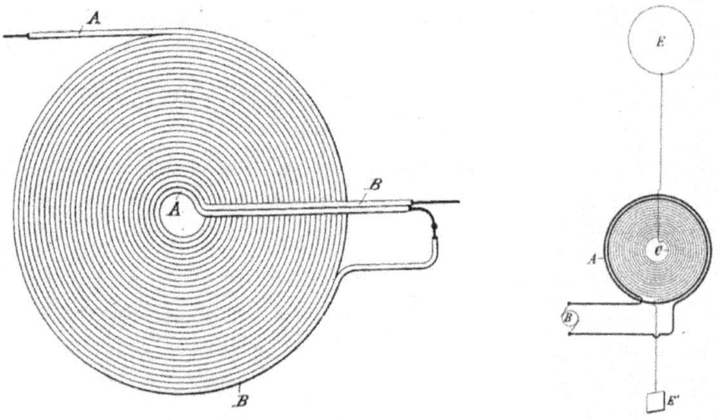

Figure 2.3: Left: Tesla's bifilar coil. [1] Right: One of Tesla's original "Tesla coils," consisting of an impulse generator (B), two flat coils (A and C), a ground connection (E') and a spherical antenna (E).[2]

The proper use of impulses allowed for the generation of Radiant Energy: shock waves emitted in the surrounding field that could move right through the air and other materials with little to no loss of power. As a result of these shock waves, in addition to the increase in voltage, there was an increase in electric current. This is crucial since the electrical power of a device is the product of its voltage and current.

Tesla amplified the power in his coil by several thousand times the power needed to drive the device. This made him one of the very first pioneers in what we now call free energy.

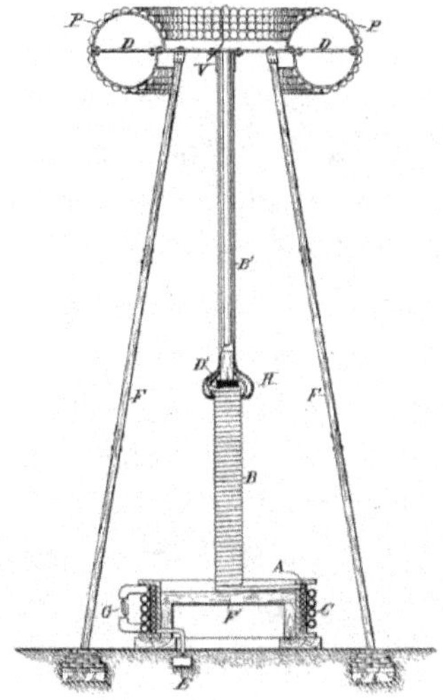

Figure 2.4: The Magnifying Transmitter from Tesla's patent "Apparatus for transmitting electrical energy."[3] The circuit uses three coils (A, B, C) instead of two.

The Magnifying Transmitter

Tesla could now generate an abundance of energy. The next challenge was to be able to transmit it wirelessly to deliver it to the consumer. Again, there seemed to be no limit to the possibilities and applications for such an invention. The next version of the Tesla coil was to be a device that could send abundant energy anywhere on the planet. He called this

his *Magnifying Transmitter*, which amplified the energy the device needed to function. This made surplus energy available to transmit wirelessly. As with his earlier versions of the Tesla coil, additional energy came from the surrounding field (the ether), set in motion by impulse electricity. Unfortunately, Tesla does not explain in his patents exactly how that energy flowed into the device, but he does point out that the energy was amplified thousands of times. As a result, the electrical power could reach hundreds of thousands of horsepower, enough to power more than ten thousand homes.

Radiant Energy through the Earth

The fact that Radiant Energy passed through all kinds of material was both an advantage and a disadvantage for its transmission. Ordinary radio waves – which had concurrently just been discovered in Tesla's time – are reflected at the top of the Earth's atmosphere and can thus easily travel long distances across the globe, far over the horizon, albeit with relatively much loss of energy. Radiant Energy, however, is not stopped by the atmosphere. Tesla, therefore, had to come up with another way to make Radiant Energy available worldwide. If Radiant Energy could not be transmitted around the globe, it had to pass through it. Tesla wanted to use the Earth as an intermediate station for both receiving larger amounts of energy and being able to transmit it to any point on its surface.

He discovered that with specific electrical vibrations from his Magnifying Transmitter, the Earth no longer acted as an electrical insulator but rather as a conductor. This required quite a lot of energy, which he had now obtained through the amplifying effect of the Tesla coil. By sending the right frequency of vibrations into the ground, Tesla could "strike" the Earth like a giant bell. The vibrations generated in this way would travel throughout the Earth and bounce back from the other side. This created standing waves, like how a guitar string vibrates when plucked. These waves could then be received at specific points on the globe, depending on the transmitter's location. Only an antenna tuned to the correct frequency could then receive the energy.

With his Magnifying Transmitter and Radiant Energy, Tesla envisioned a future in which every device, every vehicle and thus entire cities and countries could be wirelessly powered with an abundance of energy. But many other applications, such as global wireless communication, accurate navigation and remote control were suddenly within reach. Today, we use radio waves or even cables for these applications, which seem rather primitive and inefficient in light of Tesla's inventions and consume a lot of energy on both the transmitting and receiving sides. Mobile telephones, radio, and Wi-Fi are examples of the use of radio waves.

Over time, Tesla's original Magnifying Transmitter has unfortunately become synonymous with the "modern" version of the Tesla coil, a device used primarily for entertainment because of its spectacular display of sparks. With that, the true purpose and operation of the device have fallen into obscurity.

Radiant Energy and other forms of radiation

Tesla's research on Radiant Energy around 1895 also resulted in other new forms of radiation, including one that could *partially* pass through the body. With proper tuning, his Tesla coil, combined with a special vacuum tube, could accelerate electrons so that they collided with solid matter, thereby generating this kind of radiation.

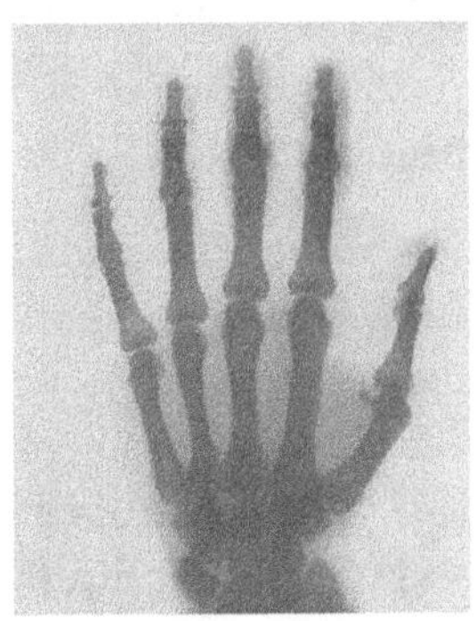

Figure 2.5: Tesla's hand,
one of the very first X-rays (1896).

We now call this radiation *X-ray*, the "X" referring to the then-unknown form of radiation investigated by Wilhelm Röntgen, who received the Nobel Prize for its discovery. Tesla, however, was a few weeks ahead of Röntgen's announcement when he accidentally made one of the very first X-ray photographs with his setup. Nevertheless, Tesla still gave Röntgen credit since he did not know the effects of the radiation at the time. On the other hand, Röntgen was very impressed with Tesla, who could soon take much better X-ray images than himself. However, X-rays were only one of the effects that Radiant Energy could produce.

Radioactivity

X-rays were discovered around 1900 almost simultaneously with the phenomenon of radioactivity – the disintegration of large, unstable atomic nuclei, emitting small (sub)atomic particles or gamma rays. X-rays and gamma rays stem from the same phenomenon: very high-energy electromagnetic waves, invisible to the naked eye. The difference lies in a shorter wavelength in gamma rays and thus a higher energy (see Figure 3.2). Both radioactive materials and X-rays are now widely used – in addition to their use in nuclear power plants – in medical diagnostics and cancer treatment.

Currently, most radioactive materials are produced from natural radioactive sources found in the Earth's crust, such as uranium and thorium. Established science assumes that such elements were created billions of years ago in the explosion of stars.

Radiant Energy from the Cosmos

Tesla, through his experiments in Radiant Energy, had become very fascinated by the newly discovered phenomenon of radioactivity. At the time, people had little idea of why some natural minerals displayed radioactivity, but Tesla was several steps ahead of science in this regard as well. Indeed, his Radiant Energy caused some materials to emit radioactive radiation or X-rays automatically. Which radiation effect was generated depended on the frequency and duration of the impulses and the material used on which Radiant Energy impinged. Tesla concluded that there had to be a source of Radiant Energy from outside the Earth responsible for natural radioactive radiation. He called this Radiant Energy from the cosmos *cosmic rays*.

Thus, according to Tesla, certain rocks or minerals converted cosmic rays into other forms of radiation. In a later phase, this led Tesla to the idea of not generating Radiant Energy himself but simply receiving the abundant Radiant Energy from the universe and using it directly. You'll read more about that later on. He would first develop his vision to make Radiant Energy an energy source widely available worldwide.

Wardenclyffe Tower

Besides being a genius physicist, Tesla was a true engineer who could physically manifest his vision and ideas. So, to test his discoveries from the laboratory in practice, he had to think big. This eventually resulted in the construction of an enormous Magnifying Transmitter named Wardenclyffe Tower. Its building on Long Island in New York began in 1901 after Tesla received $150,000 from the famous banker J. P. Morgan.

Figure 2.6: The Wardenclyffe Tower in 1904.

The extremely wealthy banker Morgan saw Tesla as a promising competitor to Guglielmo Marconi, who had successfully sent telegraph messages over increasingly long distances using radio waves. But, by this time, Tesla had abandoned his research into these "ordinary" radio waves because it was clear that his Radiant Energy technology was superior on many fronts – a whole host of new electromagnetic effects, much less energy loss in transmission and the ability to amplify electric power.

On the other hand, Morgan was initially unaware that the tower's intended purpose went much beyond just sending messages. The Wardenclyffe Tower was to be the first transmitting station that could send energy wirelessly to anywhere in the world. Moreover, once the station was up and running, it wouldn't require an external energy source such as a coal-fired power plant. In addition to its consumption, the tower would provide an enormous power that could be transmitted through the Earth at will.

Morgan, who had already initially financed Edison's direct current industry, was determined to become America's largest electricity giant. When alternating current came out on top in the race to electrification, Morgan had to enforce Tesla's alternating current patents from George Westinghouse. Morgan put financial pressure on Westinghouse and thus obtained the patent rights to alternating current. Now, he was ready to provide all of America with a brand-new electrical grid – a massively lucrative business model.

Consequently, Tesla's vision of making energy available to everyone for free was a thorn in Morgan's side. Morgan's initial investment in Wardenclyffe Tower proved insufficient, and when he got wind of Tesla's true ambition, he decided not to give him another penny. Despite numerous requests from Tesla, he was unwilling to support him further. The project came to a standstill, and Tesla was forced to take out a mortgage using the tower as collateral to pay the debts he had incurred in the hotels where he lived. Wardenclyffe Tower was eventually demolished in 1917 by the new estate owner. Tesla was yet again facing personal bankruptcy.

He was severely disappointed that his large-scale project to supply the world with clean energy had failed, but he did not give up. He would now look for a way to make Radiant Energy useful on a smaller scale. What if he could directly harness the ubiquitous cosmic rays, the Radiant Energy of the universe, without large transmitting stations? By now, it was clear to him that such an invention threatened the electricity and fuel industry stakeholders. In addition, he knew he was misunderstood by established science. With his limited resources, he was forced to work increasingly in secrecy.

The Pierce-Arrow car

"Electrical power is everywhere present in unlimited quantities and can drive the world's machinery without the need for coal, oil or gas..."

Nikola Tesla

Tesla had lost considerable popularity after Wardenclyffe, and much less is known about his work from this latter phase of his life. Nevertheless, he always continued his fundamental research and its application in remarkable inventions. One invention from his extensive body of work after Wardenclyffe particularly stands out. It is the electric car developed by Tesla, which required no fuels and did not even carry large batteries but ran solely on "cosmic rays." Did Tesla finally succeed in generating an unlimited amount of energy where it was needed, namely in the car itself?

In the 1980s, a transcript surfaced of an interview with Peter Savo, a cousin of Tesla, conducted by aeronautical engineer Derek Ahlers.[4] In this interview, Savo, an Austrian Air Force pilot, states that he was hosted by Tesla in Buffalo, New York, in 1931. There, he would witness Tesla's latest invention, which appeared to be a standard Pierce Arrow, a then-popular luxury car produced in Buffalo. Tesla had replaced the combustion engine with an 80-horsepower electric motor and had removed the fuel tank. Inside the dashboard was an energy receiver, which included a set of resistors and 12 vacuum tubes, three of which were of type 70-L-7. Savo could specifically recall the small dimensions of the device. A two-meter antenna was attached to the car. Except for a regular 12 Volt battery, no other batteries were present. Tesla pushed two rod-like buttons on the receiver and said, "We now have power."

Figure 2.7: A restored 1930 Pierce Arrow.

Tesla gave Savo the car key for a test drive. He also said that if you connected the receiver to a residential home, the receiver would provide enough energy to power both the car and the house. Savo and Tesla then took a test drive of eighty kilometers, during which the vehicle reached speeds of up to 140 km/h. When Savo asked Tesla where all this energy came from, he replied, "It is a mysterious radiation that comes out of the ether." An inexhaustible source that he felt humans should be grateful for, as it would soon be powering all of our vehicles forever. He was otherwise quite secretive about the workings of his invention, so many of Savo's questions remained unanswered.

Tesla did tell him that he was working with a shipbuilding company to build a ship with this technology, but Savo was not allowed to know which company was involved. Finally, after eight days, during which Savo and Tesla tested the car together, Tesla left the car under strict surveillance at a remote site near Buffalo. However, he took the energy receiver from the dashboard and the car key with him. Savo has since been unable to figure out where the car or the energy receiver ended up. Even drawings of the invention have not surfaced anywhere since.

Savo's story at first sounds like a nicely crafted science-fiction tale, were it not for the fact that in 1932, on his 76th birthday, Tesla himself had the following entry recorded in the *Brooklyn Daily Eagle* newspaper:[5]

> *"I have harnessed the cosmic rays and caused them to operate a motive device. (...) All of my investigations seem to point to the conclusion that they are small particles, each carrying so a small charge that we are justified in calling them neutrons. They move with great velocity, exceeding that of light. More than 25 years ago I began my efforts to harness the cosmic rays and I can now state that I have succeeded in operating a motive device by means of them. (...) I will tell you in the most general way [that] the cosmic ray ionizes the air, setting free many charges – ions and electrons. These charges are captured in a condenser which is made to discharge through the circuit of the motor. I have hopes of building my motor on a large scale, but circumstances have not been favorable to carrying out my plan."*

It seems plausible that Tesla succeeded in making cosmic Radiant Energy useful on-site, but that has never been officially confirmed. We're unsure if Tesla means the Pierce Arrow from Savo's account by his cryptic "motive device." In any case, Tesla spent the last phase of his life mainly on this cosmic radiation, the Radiant Energy that abounds everywhere and at all times. His use of vacuum tubes to receive or amplify cosmic rays would not be surprising either since Tesla was known to produce Radiant Energy effects using vacuum tubes. It is up to all interested scientists and engineers to further the research on Tesla's energy technologies.

Tesla's legacy

Tesla has held a birthday conference every year since 1931, of which the report from the Brooklyn Daily Eagle quoted above is an account. On his birthday, Tesla briefed the press on his latest developments. In addition to an engine that ran on cosmic rays, Tesla had told them that he had developed a mechanical oscillator that could cause earthquakes and even a powerful superweapon that should abolish war – Tesla's infamous "Death Ray." These inventions likely had to do with his research into Radiant Energy.

Tesla lived in the Hotel New Yorker in the last phase of his life. His hotel bill was paid by the company of his close friend and business partner, George Westinghouse. Tesla died alone in his hotel room at 86 in 1943, virtually broke and without the recognition he deserved. Two days later, all of his possessions were seized by order of the FBI (Federal Bureau of Investigation). Remarkably, one of the investigators of Tesla's estate was John G. Trump, a professor at MIT (Massachusetts Institute of Technology) and the uncle of the 45th president of the United States, Donald J. Trump. It was up to John Trump to determine whether Tesla's inventions could pose a threat to national security the moment they fell into the wrong hands. Of course, the U.S. government was interested in them, especially new weapons and energy technology. Significantly, of the 80 crates of material, only 60 were ultimately returned to Tesla's family. This happened after John Trump officially concluded that the material "did not include new, sound, workable principles or methods." Unfortunately, we can only speculate about the withheld information. The fact is that this mystery has spurred entire generations of researchers to discover what the U.S. government found interesting enough to keep secret.

Based on the tremendous success of Tesla's numerous inventions shared with humanity, we may assume that his insights on Radiant Energy will also hold true. Meanwhile, we have become aware of the resistance to the roll-out of such technology among certain stakeholders. With what Tesla had said about Radiant Energy before his death, humanity must try to reinvent the wheel itself, and luckily, there are many examples of this.

In that respect, Tesla's spirit visibly lives on to this day. Not only because we use his inventions every day as soon as we deal with electricity, something that many people – including scientists – do not realize, but his vision of providing the world with unlimited clean energy also lives on. Countless researchers followed in his footsteps. Thomas Henry Moray, Edwin Gray,[6] Konstantin Meyl[7] and Eric Dollard[8] are just a few scientists and inventors who managed to reproduce parts of Tesla's most crucial technologies. We will elaborate on this in the next chapter.

Tesla was ahead of his time, and through his contributions, he is rightly considered both the forerunner of the electric age and the founder of the free energy movement.

1 Tesla, N. (1894). Coil for electro-magnets (U.S. Patent No. 512,340).
2 Tesla, N. (1905). Art of transmitting electrical energy through the natural mediums (U.S. Patent No. 787,412).
3 Tesla, N. (1914). Apparatus for transmitting electrical energy (U.S. Patent No. 649,621).
4 teslauniverse.com/nikola-tesla/articles/nikola-teslas-electric-car-folklore-or-historical-fact
5 John J. A. O'Neill, "Tesla Cosmic Ray Motor May Transmit Power Round Earth," Brooklyn Daily Eagle, July 10, 1932.
www.newspapers.com/image/693410893/
6 www.rexresearch.com/evgray/1gray.htm
7 www.meyl.eu/go/indexb830.html
8 ericpdollard.com/

Chapter 3:
Where does free energy come from?

Throughout history, there have been many inventors and scientists who, like Tesla, have found a unique way to generate unlimited clean energy. All these inventions have one thing in common: they draw energy from a source that can be tapped anywhere and anytime. The name for this inexhaustible source used for thousands of years is the ether, but it has also been given numerous other titles. In this chapter, we'll further explore this infinite sea of energy and how free energy technology can tap into it – energy that permeates and moves everything in the universe.

What is the ether?

For as long as humankind can think, he has been trying to understand how everything works, why everything is the way it is, but above all, who he is himself and what he is doing here. Gathering knowledge is not reserved for philosophers or science. Every human being seeks understanding and meaning in their unique way.

In recent decades, science has been searching for a "theory of everything" that brings all the laws of nature under one umbrella. How can all facets of our reality be connected? For thousands of years, countless cultures worldwide have argued that there can ultimately be only one source. For creationists, that source is the Creator, God, the One; for science, that source – the first moment of creation – is the Big Bang. This book will not dive into theology or ontology too deeply, but let's assume that whatever the source or origin, our universe "runs" on what we call *energy*. Everything *is* energy in one form or another, whether it is matter or light, visible or invisible. If we reduce all these different forms of energy to a still undefined "primordial soup," we are speaking of only a single source of yet *unmanifested energy* – energy not yet in the form of atoms or light.

This one energy source is thus at the deepest root of our physical existence – visible and invisible energy and mass. Since ancient times, this source has been called *the ether*. An energy field that connects and nourishes everything: every atom, every molecule, every human and every star in the universe. The ether itself, however, is usually invisible and unmeasurable to us – and also to modern science – because it resides, as it were, in the *background of reality*. Our measuring devices and the underlying accepted science cannot yet measure the unmanifest directly. Yet this field explains much of what current physics does not yet understand. For example, how can a light wave propagate through so-called empty space? Where does an electron get its energy to spin around the nucleus of an atom indefinitely? Why does the universe seem to be expanding at an accelerating rate?

The latest scientific discoveries confirm that empty space, the vacuum, is never really empty but rather full of energy in its unmanifested state. Thus, science is returning to the idea of the ether after the concept was out of fashion for a long time, as we will see further on.

Understanding the ether can not only solve some of the mysteries of our existence – the ether can, above all, serve as an inexhaustible source of energy in the form of electricity and heat. Energy we need to build a fairer and cleaner society – on which we will elaborate later. First, we look at how the concept of the ether originated, why it went out of fashion and how it is making its resurgence.

From the Vedas to quantum physics

"Ere many generations pass, our machinery will be driven by a power obtainable at any point in the universe."

Nikola Tesla

Akasha

More than 10,000 years ago, ancient Indian culture had a term for an element that fills all of space and is essential to the existence of other elements. It was the first and most important element and thus formed the basis for everything we call "physical." In the Upanishads, the last volume of the oldest and holiest scriptures of India – the Vedas – this element was called *akasha*. Akasha can be translated from Sanskrit as "open space" or "void." As one of the foremost free energy pioneers,

Nikola Tesla was deeply inspired by the philosophy of the Vedas. In his 1907 article "*Man's Greatest Achievement*," he wrote:[1]

> "*Long ago [mankind] recognized that all perceptible matter comes from a primary substance, of a tenuity beyond conception, filling all space, the Akasha or luminiferous ["light-bearing"] ether, which is acted upon by the life-giving Prana or creative force, calling into existence, in never-ending cycles, all things and phenomena. The primary substance, thrown into infinitesimal whirls of prodigious velocity, becomes gross matter; the force subsiding, the motion ceases and matter disappears, reverting to the primary substance.*"

Tesla also referred to *prana*, Sanskrit for "breath of life" or "vital energy." The Vedas thus describe how everything in the world results from a conscious life force (prana) interacting with the universe's infinite, as yet unmanifested energy (akasha). Prana "resides" in everything that lives – it is the essence of life. Other Eastern philosophies in Japan and China, for example, call this life energy Chi or Qi. Life energy can be used directly through disciplines such as Tai Chi and Qi Gong to keep the body healthy and strong. In holistic medicine techniques such as Reiki, life energy is used to heal the body at its most fundamental, energetic level.

The term akasha is also used to indicate a field in which the entire history and knowledge of the Earth, humanity, and the whole universe are stored: the *Akashic Records*. By spiritually connecting with this "primal library," one has access to all the knowledge and wisdom that exists and is appropriate at that time. Many influential scientists have stated that their most significant breakthroughs came to them from such an infinite field of knowledge as sparks of inspiration. Isaac Newton is a prime example. He was an alchemist at heart who engaged in esoteric science and the expansion of consciousness *before* he laid the foundation for modern physics. Nikola Tesla, Albert Einstein, and renowned quantum physicists such as Erwin Schrödinger also declare to have drawn their inspiration from such a source.

The akasha is thus not only the source of everything physical but also contains all the information for everything *non-physical*: thoughts, emotions and ideas. In accordance, the term *information* – from Latin: *in formare* – means "to put into form" the yet unmanifest.

The fifth element

The word *ether* – classically written *aether* – derives from the Ancient Greek *aithèr,* meaning "pure, fresh air." In Greek mythology, Aithèr is also a deity, the god of the higher air that all gods breathe. The Greek philosopher Aristotle, a student of Plato, described, in addition to the four primary elements we all know – earth, water, air and fire – a fifth element that was even more fundamental. Unlike the other elements, this one was not subject to time but constantly in a local, swirling motion and present everywhere. Later, this element was integrated into the system of elements as the *fifth element.* Today, we still see this concept in the word *quintessence,* which literally means "fifth essence." The term denotes that which it is all about. Aristotle used the notion of the ether to explain how the other elements relate to each other and how space was filled with it so that celestial bodies could make a perfect circular motion. In Greek philosophy, the ether, like the Vedic akasha, is the most fundamental substance from which the universe arises.

Ether in modern history

During the Middle Ages, the system of the five elements spread throughout Europe. The ether or quintessence was an essential concept for alchemists – including Isaac Newton – who were the top scientists of their time. In alchemy, they tried to capture the ether and use it in medicines and potions that would have powerful healing effects due to the ether's purity.

Around 1700, the ether was also integrated into emerging natural science when Isaac Newton laid the foundation for what is now called classical physics. In his theory of how light works, he hypothesized that the ether was the light-bearing medium that allowed light waves to travel, just as waves travel across the ocean. Without water in the ocean, no waves could ever reach the shore. In other words, without the ether, we wouldn't see anything because sunlight would never reach our eyes. Newton also described the characteristics of the ether in terms of density. Where the ether has the highest density, light is most strongly refracted. This was his explanation for the observation that light is refracted more strongly by glass and water than by air. After all, the density of the ether coincides with the density of atoms and molecules. It even applied to Newton's formulation of gravity: the closer the atoms were together, the higher the ether density. Gravity could be seen as a force of *pressure* from a lower to a higher ether density.

However, toward the end of the 19th century, the ether concept went into dire straits. Physicists Albert Michelson and Edward Morley devised an experiment to test whether the Earth actually moved through a stationary ether. If the stationary ether hypothesis were true, they should be able to measure an "ether wind" blowing past the Earth as it moves around the Sun – there should be some friction between the Earth and the ether. By splitting a beam of light into two rays and recombining them in a so-called interferometer, the scientists expected that one ray could be visibly slowed down by the ether wind relative to the other. However, no such effect was measured. Instead, both light rays were consistently equally fast, regardless of the direction in which the beam was split. They concluded that, therefore, the ether could not exist. In doing so, however, they assumed that the ether was *stationary* relative to the moving Earth – an assumption they didn't put to the test.

The damage was already done. Without much further research, established science quickly rejected the idea of the ether. For the first time in many centuries, the vacuum of space was again seen as genuinely empty. With the Michelson-Morley experiment in mind, Albert Einstein formulated his special theory of relativity, where the ether no longer appeared in the equations. This then struck the fatal blow for most physicists. The concept of ether could be thrown out, despite Einstein himself declaring that this conclusion was drawn too prematurely. After all, his theory of relativity still required an ambient medium with specific properties. Therefore, empty space could not be *truly* empty.

Development of quantum physics

Early in the previous century, next to Einstein's development of the theory of relativity, an entirely new field of science was discovered:

quantum physics – the study of particles smaller than an atom, such as electrons and protons. To this day, scientists are still trying to unify quantum physics with the worldview of Newton's classical physics. What are these quantum particles made of, and how do they interact with each other and the world we know? Quantum physics shows us that, at the smallest level, a particle cannot have a precise position or velocity but only forms a *probability distribution* – a probability of finding a particle in a specific location. In fact, all particles are simultaneously *waves* spread out over infinite space. It is energy that manifests either as a particle or a wave, depending on how we measure it. These findings alone have significant implications for our classical picture of reality. But quantum physics also forces us to reconsider the idea of "empty space."

Zero-point energy

If we would isolate an atom and try to cool it to the absolute zero in temperature – zero degrees on the Kelvin scale, corresponding to minus 273 degrees on the Celsius scale – it turns out the atom still retains a small amount of kinetic energy. Many scholars did not expect this. After all, temperature is a measure of the kinetic energy of matter: the higher the temperature, the faster atoms move around and the more often they collide. In theory, the motion of all matter would cease at absolute zero, right down to the smallest subatomic particles. But no matter how hard we try to rid the atom of all its kinetic energy, it continues to vibrate. Therefore, it must continually receive energy from somewhere to keep the atomic nucleus vibrating permanently and to keep the electrons moving at high speed around the nucleus. In that respect, we have never observed zero degrees Kelvin – matter is never really motionless. This is why scientists call this ground state of energy *zero-point energy*. It is the state of minimum energy of a particle or atom at the temperature of absolute zero. But where does this energy come from? Does it come from the vacuum, seemingly empty space? It sure begins to look like it. That's why zero-point energy is also called *vacuum energy*.

In 1948, Dutch physicist Hendrik Casimir devised a relatively simple experiment to investigate this zero-point energy. In this experiment, he envisioned two thin mirrors in a vacuum placed very close together. His hypothesis was just as simple: if space contains zero-point energy in the form of waves – "fluctuations" of the vacuum – then fewer *modes* of these waves fit into the space between the mirrors than in the space outside of them. As a result, the pressure of the waves in the area outside the mirrors would be greater than between them, and the mirrors would be pressed toward each other. And indeed, this is precisely what scientists observed following Casimir's prediction. We have since called this the *Casimir effect*. It is one of the proofs for "empty space" not being empty at all but filled with the zero-point energy field, which can exert force.

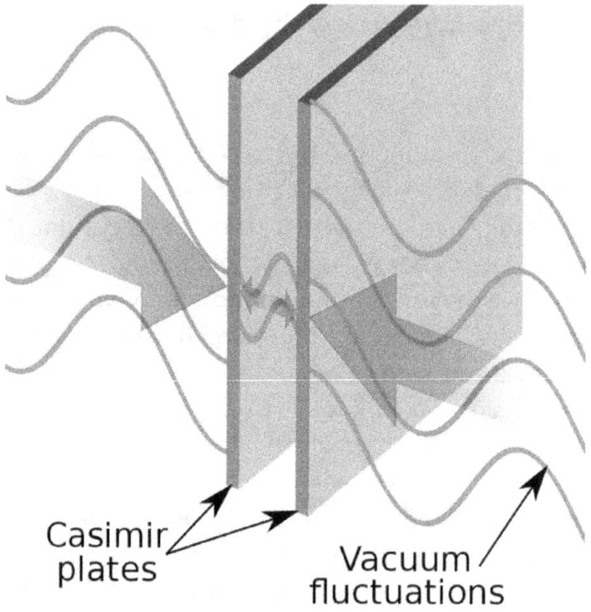

Figure 3.1: The Casimir effect. There are fewer vacuum fluctuations (grey waves) between the mirrors than outside them, pushing them toward each other.

Today, zero-point energy is used to explain several mysteries in physics. For example, zero-point energy explains why electrons around an atomic nucleus can spontaneously change orbit, and it is also associated with the universe's accelerating expansion. Indeed, this acceleration must cost unimaginable amounts of energy – a curious phenomenon attributed to what scientists call "dark energy," a yet unknown, invisible energy that appears to comprise as much as 72% of all energy in the universe. Notably, less than 5% of all energy in the universe is a form of energy (light and atoms) that current science can measure and describe.

Other names commonly used in science for this mysterious energy are "cosmological constant," "quantum foam," and "quintessence."

Now, it becomes quite clear that the concept of the ether was not so loony after all. Space is filled to the brim with energy that exerts influence from the smallest quantum scale to the largest galactic scale.

So how much energy is there in this ether, which permeates all of space and everything in it? Scientists are still in disagreement. So far, only estimates have been made. Paul Dirac, one of the founders of quantum physics, suggested that the energy in the vacuum should be gigantic. In his view, the ether – which received the name "the Sea of Dirac" in his honor – is an infinite sea of negative electrical charge. Experts in Einstein's theory of general relativity argue, on the contrary, that the energy must be small enough for that theory to be consistent with astronomical observations. It must be noted that most physicists avoid the concept of infinity altogether because it cannot be used in practical mathematical equations.

Two well-known scientists, John Wheeler and Richard Feynman, suggested there is enough energy in the volume of a teacup to boil and completely vaporize *all the world's oceans*. Converted into electricity, this would be enough energy to power the entire planet for over a million years!

Now, imagine if we could put this energy to practical use. Energy would become so abundant that it would eliminate the need for solar panels, wind turbines and fossil fuels altogether. The enormous positive impact on our global society would be hard to fathom. The question then becomes: How do we tap into this source of abundant energy?

"*If you want to find the secrets of the universe, think in terms of energy, frequency and vibration.*"

Nikola Tesla

Free energy:
Tapping into the source of life energy

An immeasurable sea of energy surrounds us like water surrounds fish in the ocean. In principle, everything consists of vibrating energy, waves that manifest from the ether as our tangible reality. So, how do we convert this all-pervading energy into a form that is useful to us and, at the same time, does not disturb everything else? Let us first consider some essential concepts.

Fields

A *field* can be seen as a medium of a certain quantity, extending across space. Temperature, for example, can theoretically be measured at any point in the universe and thus be mapped into a gigantic "temperature field." Such a field, with a value of a certain quantity at each point (e.g., temperature or pressure), is called a scalar field. When a field not only has a value at each point but also a direction, it is called a vector field. If you draw a line through several points in a vector field following the path of each vector, you are drawing a field line. The weatherman's wind map is an example of a vector field, where wind speed and direction are given at each point on the map. But we also have electric fields, magnetic fields and the combination of both – electromagnetic fields – which have both a strength and a direction.

Particles, waves and fields

All matter consists of what current science calls *particles*. Molecules are made up of atoms; atoms are made up of protons, neutrons and electrons. These, in turn, consist of even smaller particles. As you have already seen, quantum physics teaches us that all particles are at the same time *waves*. Therefore, particles do not have a precisely defined position or speed but are small "packets" of vibrating energy. How we measure or observe this vibrating energy determines whether it manifests as a *particle* or a *wave* at that moment. Particles in their wave form are like waves on the ocean's surface. But if water is the medium that waves on the ocean, what medium is it that "waves" tiny particles like electrons? We can call this medium the field of ether.

Every particle is a wave, and that which "waves" is a field. You might as well say that everything consists of fields since all forms of energy consist of vibrations in fields that extend to infinity in all directions. In this way, we can see that everything is connected. Everything is vibration in fields, fields of vibration.

Two types of waves

We can distinguish two wave types that differ in how the medium (the field) supports the movement of the wave. *Transverse waves* protrude from the ocean's surface, with the medium (water) moving up and down. The motion of the medium – up and down – is perpendicular to the motion of the wave. When thinking of "waves," it is this type that most people visualize; the shape of these waves is very recognizable. Other examples of transverse waves are light waves, which can be seen with the naked eye, and telecommunication waves, which cannot be measured without special devices. Yet these examples are all part of the same *spectrum of electromagnetic radiation*, which also includes infrared, microwave and X-ray radiation.

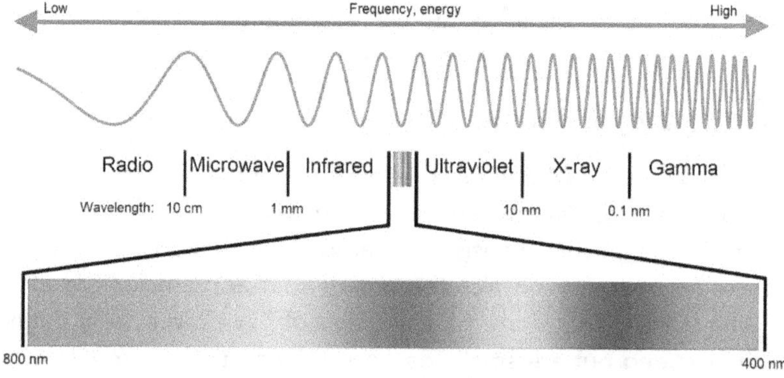

Figure 3.2: The spectrum of transverse electromagnetic waves. The shorter the wavelength, the higher the frequency and the higher the energy of the waves. The light we can see – all the colors of the rainbow – is only a tiny part of the spectrum.

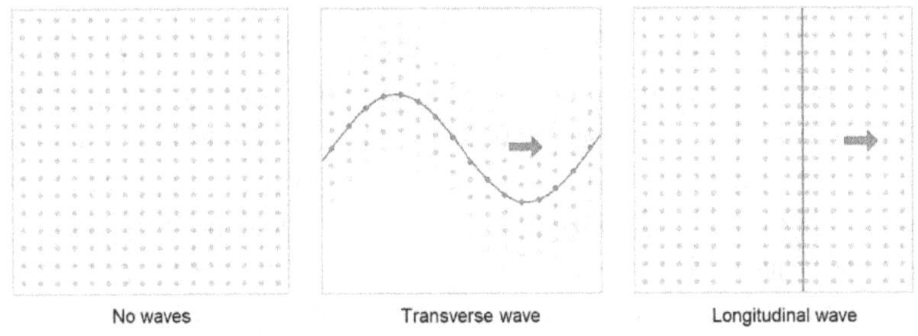

Figure 3.3: Two types of waves. On the left: field without waves. The field points are stationary. Middle: transverse wave. The field points (or particles) move up and down, perpendicular to the motion of the wave (arrow). Right: longitudinal wave. The field points (or particles) move back and forth in the same direction as the motion of the wave.

Much less known to most people are so-called *longitudinal waves*, in which the medium does not move up and down but alternatingly compresses and expands. Sound moving through the air is a prime example of this. Sound consists of regions of moving air molecules of alternating high and low density – areas of high and low pressure. Here, the medium does not move up and down but "back and forth" in the same direction of travel as the wave.

If we compare these two wave types in one medium, longitudinal waves generally travel faster than transverse waves. This is true, for example, in earthquakes, where the first (longitudinal) shock wave passes through the Earth's crust about twice as fast as the transverse wave that follows. What could this mean for the medium of the ether, in which transverse electromagnetic waves move at the "maximum speed limit" – the speed of light?

Looking at the theories behind Tesla's Radiant Energy and similar technologies, we often come across inventors using a new mode of electromagnetic (EM) waves, namely longitudinal EM waves. "Ordinary" light waves (transverse waves) move with the speed of light, but it is believed that longitudinal EM waves can travel much faster than light and pass through all kinds of materials.

Resonance

Resonance is a fundamental concept in physics. Simply put, resonance is the vibration of different waves at the same frequency, making them amplify each other. Here, the frequency – how often a wave goes up and down or back and forth in time – is crucial. Two waves can only resonate properly if they have the same frequency. Imagine a singer who can break a wine glass with her clear voice. When the glass is tapped, a pure "ping" sound of a particular pitch or frequency is heard. If the singer sings at exactly that pitch near the glass, the glass will vibrate accordingly by resonance. The vibrations of the glass and the sound of the voice reinforce each other until the glass can no longer bear the vibration and breaks.

Another example is a swing. If someone on the swing already moves back and forth a little, and you push on one side at the right moment (with the same frequency as the swing motion), it takes minimal effort to make the swing motion very large. In this way, any wave, be it sound, light, or any other type of vibration, can begin to resonate.

The idea behind free energy is that we come into resonance with vibrations in the energy field of the ether so that the ether can provide our systems with energy.

Entropy and syntropy

Entropy can be thought of as the tendency of energy to spread out as much as possible so that there is an equal amount of energy everywhere. We deal with entropy every day. When we brew coffee and pour it into our cup, it gradually cools down. If we do not drink the coffee, we can be sure that the coffee will eventually cool down to room temperature. We won't see a cup of cold coffee spontaneously heat up again. This is not impossible, strictly speaking, but the probability of that happening is astronomically small. This one-way traffic is due to the physical phenomenon of entropy. The energy in the molecules of coffee, in the form of heat (kinetic energy), tends to spread out as much as possible so that there is no longer any difference between the energetic state of the coffee and that of its surroundings.

But this is only half of the story. How can people, animals and plants grow if energy only tends to settle with the environment?

Entropy also has to do with the degree of the *ordering* of energy. If we mix a tray of red marbles with a tray of blue marbles, we end up with a tray of marbles where the original "order" of only red and only blue has vanished. It takes quite a bit of effort to reorder all the marbles. This re-ordering of the marbles – and energy in general – is the inverse of entropy and is called *syntropy* (or *negentropy*, from *negative entropy*). Syntropy is the process of the creation of all matter and all life. It is the process that creates order out of the disorder of the endless energy field of the ether. A beautiful analogy for this is a tree. A tree begins as a tiny seed that draws energy from its environment through water, nutrients, sunlight, and air to grow bigger and bigger. The tree is an ordering of the energy of its environment, which would otherwise be relatively disordered.

With the ideas of entropy and syntropy, we can also better understand our current energy systems. For example, a combustion engine uses energy in the form of fossil fuel (gasoline made from crude oil) to provide useful work. In the process, we burn the hydrocarbon molecules of the gasoline, which are converted into several smaller molecules and lots of heat. The natural "arrangement" of energy is thus broken down: the energy contained in the gasoline disperses into the smaller gas molecules and heat, both of which are absorbed into the atmosphere. The tray of red and blue marbles gets mixed up again – and our cup of coffee gets cold. We thus use entropy to power our cars. The same principle applies to all coal, gas and nuclear power plants. With these systems, we destroy ordered forms of energy to perform work, but we don't reverse the disorder we create. It is a one-way transaction.

Closed and open systems

So, what about wind turbines and solar panels? Let's look at a solar panel as a system in and of itself. The panel draws energy from the environment – sunlight – and "organizes" that energy into electricity. Wind turbines do the same but use wind as the energy source. We can see that they are indeed systems based on syntropy. It is also immediately apparent that solar panels and wind turbines are *open systems*. There is an unrestricted inflow of energy from the environment, which the system organizes into useful energy. Here, we deliberately ignore all it takes to produce these devices.

However, what we define as "the system" is essential here. A combustion engine with a gasoline tank can be considered a *closed system* because the energy source is included in the system. The amount of gasoline is limited, and when it runs out, the engine stops. But in reality, closed systems do not exist. A closed system is only an imaginary box we place around the system. Earlier, we noted that all matter and all other forms of energy are constantly connected to the ether. Every atom – and thus every cell in your body, every tree and every planet – is continuously connected to it. The entire universe is an open system energetically sustained by the ether.

Balance and the toroidal energy field

We could also compare entropy to an *explosive* and syntropy to an *implosive* tendency. A significant shift is currently taking place in our ways of generating energy as we move from explosion-based energy generation to implosion-based systems. This is a critical move because explosion in itself is a destructive, life-destroying force that creates disorder, while an implosion is a constructive, life-generating force that brings order. Implosion is nature's way of bringing energy together so that life can grow. Explosion is nature's way of allowing things to decay and unite with the environment to be transformed back into new life. This completes the cycle. In nature, explosion and implosion, entropy and syntropy, order and disorder, are in balance.

This balance is expressed energetically in an energy field shaped like a donut. Such a three-dimensional donut shape is called a *toroid*; thus, we call the corresponding energy field the *toroidal field*. It can be seen in many places in nature. Let's retake the example of the tree. At the bottom of the toroid, where the tree roots stick into the ground, syntropy (implosion) occurs where water and nutrients are sucked upward. Next, the tree brings the energy together at the toroid's center, the tree trunk. Here, all the building blocks are put together – syntropy causes the growth of the tree trunk, branches and leaves. Then entropy (explosion) causes the leaves to fall occasionally and decay into the ground, feeding the soil and bringing the toroidal shape of energy back to a full circle.

The characteristic shape of the toroidal energy field is also seen in seeds, fruits, flowers and animals. Humans, too, have a magnetic field around them with this shape centered in the heart, as does every atom, planet, and star. The toroidal field is the universe's way of creating a balance between order and disorder, yang and yin, on every conceivable scale. The more we integrate this balance into our systems and technologies, the more we harmonize with nature.

Figure 3.4: Examples of toroidal geometry in the fields of nature. Left: the magnetic field of the human heart. Center: cross-section of an apple. Right: the magnetic field of the Earth.

The vortex

Many free energy pioneers have indicated that they have found a way to create a toroidal field that causes the ether to implode and transform into another form of energy. The ether is considered a fluid-like medium like water, but thinner, yet less compressible (stiffer). It is believed that an excellent way to make ether implode somewhere is to create a *vortex* (plural: *vortices*). This can be demonstrated well when a full bathtub is drained. A spiral whirlpool is automatically created in the water to make the water flow through the drain as smoothly and efficiently as possible. If the vortex moves inward, then energy converges – implosion takes place. A vortex moving outward does the opposite. A toroidal field consists of two vortices: one moving inward and one moving outward. Vortices also occur everywhere in nature: we see them in hurricanes, whirlpools in the water of flowing rivers, eruptions on the Sun's surface and even in the shape of our Milky Way galaxy.

Coefficient of Performance (COP) and over-unity

Most energy technologies we use today are less efficient than we would like. As a result, we are facing significant energy losses. To illustrate, an internal combustion engine in a car converts only 10 to 40 percent of the energy available in the fuel into useful energy to power the vehicle. The rest of that energy is wasted as heat and exhaust fumes. And we're not even considering the energy required to extract petroleum, produce the gasoline and get it into the fuel tank.

An electric motor is not perfect either: about 60 to 90 percent of the electrical energy can be converted to useful energy, and the rest is heat. Again, we are not considering what it takes to produce the electrical energy in the first place.

Energy efficiency can also be expressed as a *coefficient of performance* (COP), which indicates the ratio of useful energy you get out of the system to the energy you must put in. So, for a combustion engine, the COP is between 0.1 and 0.4.

However, it is possible to make systems with a COP greater than 1. In that case, we can speak of *over-unity*, where the system produces more useful energy than it consumes. But, of course, that extra energy has to come from somewhere.

A classic example of a "closed system" with a COP greater than 1 is the refrigerator in your kitchen. For every unit of electrical energy the refrigerator consumes, it can move three units of heat from the inside to the outside. This gives a COP of 3, but that number is somewhat misleading. This is because we cannot put the excess efficiency to good use at the end – the fridge is cooled down at the "cost" of a slightly heated kitchen. In a way, all the energy we put into the refrigerator is "lost" in cooling the inside and heating the space around it. There is no additional energy input that we can make use of. Therefore, we cannot cool down the kitchen by leaving the refrigerator door open because net energy is lost to heat.

But what if we could build an open system with a COP greater than 1 – working in a state of over-unity – drawing in excess energy from the ether?

Perpetual motion machines

Current physics assumes that energy cannot be created out of nothing, nor can it be destroyed. Instead, energy can only be transformed from one form to another, for example, from kinetic energy to heat. These principles of the transformation of energy are described by the *laws of thermodynamics*.

The standard interpretation of these laws is that you can never build a motor that runs forever because a motor always loses energy to friction and thus heat. Such an eternally running engine is called a *perpetual motion machine* and is often represented as a wheel driven continuously by gravity, with one half of the wheel mechanically made heavier than the other half. But, unfortunately, this does not work: 'dropping' the heavier half of the wheel always takes the same amount of energy as 'lifting' the lighter half, and friction causes the system to lose energy over time.

Figure 3.5: A "perpetual motion machine" in the form of a gravity wheel. The ball bearings fall farther off the axis on the left side than on the right, hoping that the wheel will turn counter-clockwise forever – but unfortunately, that doesn't work.

However, these principles apply only to a closed system in which the environment supplies no additional energy. Moreover, we find perpetual motion in every atom: the particles in an atom vibrate and spin eternally. So the question is not whether perpetual motion exists – because it is everywhere in and around us – but the question is how we can tap into the source of this perpetual motion to use it directly.

Free energy technology

We now come to a description of the possible operation of free energy technology. Free energy devices are *open systems* interacting directly with the ether through resonance. They use syntropy (implosion) to turn the disordered energy of the environment into useful energy, just as a tree grows through sunlight, water and nutrients. The device creates a toroidal energy field where a vortex draws in ether energy to amplify the energy it needs to function. The device works in a state of over-unity, a COP greater than 1 – the device releases more useful energy than it consumes. That useful energy can be electricity, motion or heat. Some of that energy can be returned to make the device self-powering, even eliminating the need for a common external energy source.

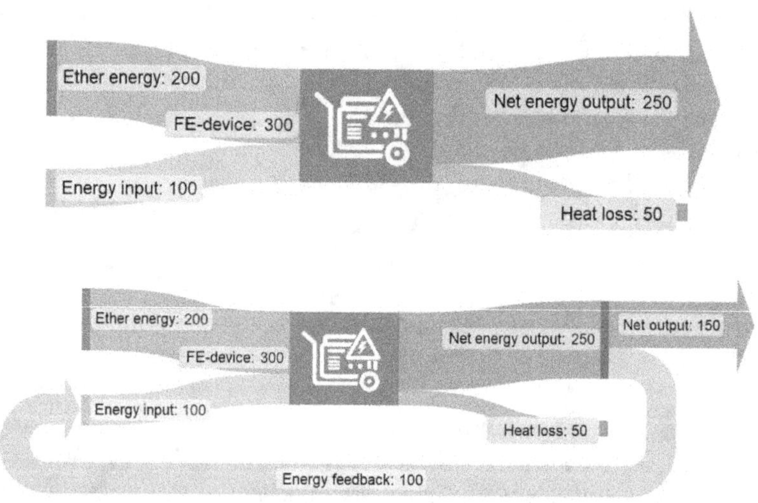

Figure 3.6: Above: Schematic representation of energy flow in a free energy device. The amount of energy that goes into the system equals the amount of energy that comes out. Below: If part of the energy output is fed back into the system, the device is self-powering and does not need an external energy input.

The ability to heal

In principle, every living being is a self-contained system remarkably capable of healing itself. Often, in today's medicine, we are looking for external remedies and interventions to make us healthy, while the human body has so much within it to stay healthy and strong. The placebo effect is underrated evidence that we can also harness free energy in our bodies. We are connected to the source of life energy and can heal ourselves by consciously aligning with this source. There are numerous good sources of information about this, so we will not explore it further here. For example, read *The Biology of Belief* by Bruce Lipton.[2]

That said, there are ways to support our bodies by connecting with nature and technology in harmony with nature. Since free energy uses a constructive force that underlies the creation of all life, some free energy technologies can heal people, animals and plants. An excellent example in this field is Dr. Wilhelm Reich,[3] an Austrian physician and psychotherapist who called the vital life force *orgone* (after the word "orgasm"). With his inventions that utilized this energy, he reportedly cured his patients of various ailments, including cancer. It is still being researched today. Other examples include using pyramid structures[4] for healing and improved plant growth, using plasma to purify water and make it energetically beneficial,[5] and countless others.

Anti-gravity

Besides generating unlimited clean energy and the ability to heal our bodies and nature, free energy offers yet more keys to new areas of invention. Several free energy inventors propose that their device also affects gravity. This is not a strange idea, considering that the ether might be responsible for maintaining all particles' (apparent) mass and, thus, gravity.[6] If we locally harness the ether to generate electricity, it could be possible to locally divert the ether from sustaining gravity.

Indications emerged from various quarters of formerly secret government projects (so-called *black projects*) supporting this hypothesis. Research has been conducted for decades on how unknown flying objects (UFOs) hurtle through our skies at tremendous speeds without ever making a sound or leaving a trail of exhaust fumes behind them. News of such projects has been rampant in recent years after the U.S. government admitted in 2017 to studying the phenomenon extensively to determine whether it poses a threat.[7] [8] Beyond the pressing questions "Where do these objects come from?" and "What is their purpose?" it is also interesting to explore how such vehicles are powered to effortlessly eliminate gravity and perform seemingly impossible maneuvers, at least according to established laws of physics. Likely, the answer to the latter question is closely linked to free energy technology.

The implications of this are enormous. It implies that such technologies – man-made or extra-terrestrial – already exist and are many times more advanced than most people can imagine and could bring about a giant leap in our evolution. If you want to know more about what the UFO phenomenon has to do with free energy, read, for example, the Dutch-language books by Dr. Coen Vermeeren[9] and check out Dr. Steven Greer's movement[10] and his various documentaries. Examples of renowned inventors and scientists who worked with anti-gravity technology include Thomas Townsend Brown,[11] Yevgeny Podkletnov,[12] Viktor Grebennikov[13] and Otis Carr.[14]

Methods of free energy generation

Over the years, many different free energy technologies have been developed. To get a clear picture of this great variety, we can divide the technologies with the same working principle into several categories. We will highlight some of these below and attempt to describe their operation. This list is most likely incomplete: other forms of free energy may not be listed here. Examples of inventors are listed for each category. For the curious reader, it is highly recommended to research their technologies and backgrounds.

Solid-state

Solid-state implies a system *without moving parts*. More and more devices in our environment are solid-state because devices with moving parts require more maintenance and materials due to wear and tear. An excellent example of solid-state is the hard disk in today's computers.

Not so long ago, that was a spinning magnetic disk; today, it has often been replaced by the immobile solid-state disk (SSD).

In solid-state electrical devices, all physical, previously moving parts are replaced by *moving electromagnetic fields*. This is also the case with free energy technology. To tap into the ether, a solid-state free energy device creates resonant electromagnetic fields using special coils and other electronic components. Several theories exist about those fields and how they interact with the ether. Earlier, you read about the difference between longitudinal and transverse electromagnetic waves. A common view is that solid-state technologies use longitudinal waves to tap into energy from the ether. Examples of solid-state free energy pioneers include Thomas H. Moray (see Chapter 4), Nikola Tesla, Alfred M. Hubbard,[15] Floyd Sweet,[16] Thomas E. Bearden,[17] Tariel Kapanadze[18] and the company Saith Technologies.[19]

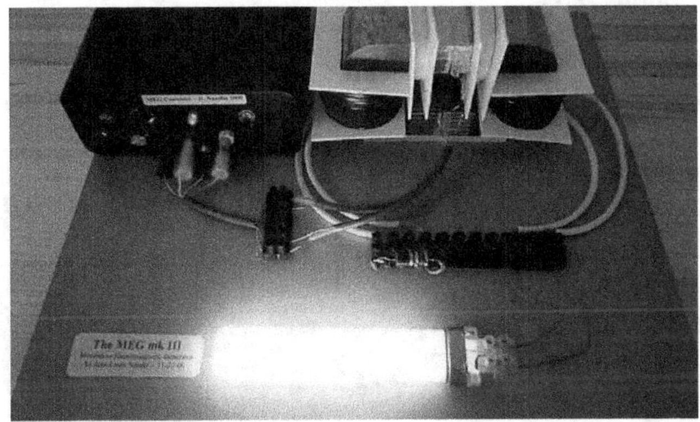

Figure 3.7: Reproduction of T. E. Bearden's solid-state Motionless Electromagnetic Generator (MEG) by Jean-Louis Naudin.[20]

Permanent magnet motor

An ordinary electric motor roughly consists of two parts: a moving rotor and a stationary stator. An electric motor converts electrical energy into kinetic energy through coils and magnets. Several people have succeeded in getting such a motor to run without a supply of electricity, using only permanent magnets – the kind of magnet that sticks to your refrigerator. By arranging the magnets in a particular way in the rotor and stator, a constant rotating force is created on the rotor greater than the friction present. Presumably, the extra energy from the ether enters through the magnetic fields of the magnets. The interaction of the rotating magnetic fields creates a vortex that draws in additional energy from the environment, creating an excess of kinetic energy. The major challenge with this concept lies in creating enough power – in addition to overcoming the ever-present friction – to tap the kinetic energy and use it to generate electricity.

Consequently, some inventors declare that the phenomenon of magnetism itself is a form of free energy since a permanent magnetic field must be constantly supplied with energy by the ether. We can, therefore, use magnets as converters of ether energy. Examples include the motors of Howard R. Johnson,[21] Muammer Yildiz[22] and the Korean company InfinitySAV.[23]

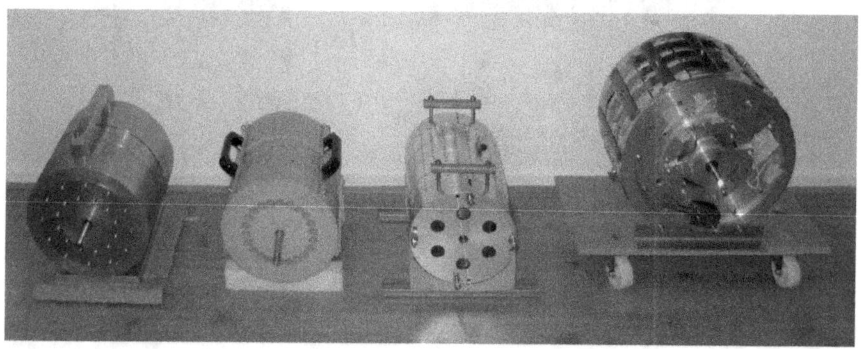

Figure 3.8: Different versions of Muammer Yildiz's permanent magnet motor. One of the motors was demonstrated in 2010 at the Delft University of Technology[24] and Eindhoven University of Technology, among others.[25]

Ether motor-generator

This category uses both permanent magnets and moving electromagnetic fields and is thus a combination of the previous two categories. It often involves a motor that uses electromagnets instead of permanent magnets. This allows the magnets to be turned on and off with great precision, again leaving a net force that keeps the rotor spinning. However, it does cost electrical energy to switch the magnets and, thus, to drive the motor. Subsequently, the rotating motion of the rotor is converted back into electrical energy; hence, the device is both a motor and a generator.

As an example of a working principle of over-unity in such a system, we can use the peak voltage created when a charged electromagnet or coil is switched off – similar to the electrical impulses that Tesla generated. These are known as spikes of *backward electromotive force* (bEMF). It is thought that such spikes are caused by the ether receiving a sudden push, bringing it into resonance with the system and restoring the imbalance by briefly pushing more energy back into the system. In ordinary devices, this energy is lost in the form of heat. However, using these peak voltages to turn the electromagnets back on, ether energy can be converted into electrical energy via rotational energy. As you can see, it is a rather complex system in which timing is crucial. Examples of ether motor-generators include the devices of Adam Trombly (see Chapter 4), Lester Hendershot,[26] John R. R. Searl,[27] Bruce E. DePalma,[28] Edwin Gray,[29] Paul Baumann,[30] Joseph W. Newman,[31] John Bedini[32] and Paramahamsa Tewari.[33]

Figure 3.9: John Bedini with his ether motor-generator that can recharge batteries indefinitely.

Plasma reactor

In physics, plasma is called the fourth state of matter, next to the other states of solid, liquid and gas. While it is less familiar to most people, it is the most common form of matter in the universe. Plasma is very much like a gas, except that the particles in a plasma all have at least one electron too many or too few, making them electrically charged. As such, they are *ions*, so plasma can be called an *ionized gas*. The electrical charge means that plasma can be strongly influenced by electromagnetic fields.

Despite its unfamiliarity, we encounter plasma a lot in everyday life. For example, the colored light of a fire comes from the plasma created during combustion. Also, every electric spark is a plasma discharge, like a lightning bolt. Our Sun – and every other star – is a big ball of luminous plasma. Plasma has all sorts of unique properties that we are only just beginning to discover. There are, for example, projects using plasma to purify water, prevent diseases and pests in crops, and heal people.

Several inventors have succeeded in using plasma to generate free energy. A so-called plasma reactor is often compared to the Sun because some form of nuclear fusion seems to occur in both systems. This releases enormous amounts of energy. Exactly how such a reactor generates over-unity is still being researched. What we do know is that it requires complicated equipment and achieves high temperatures and high voltages.

Other designs are made to convert ordinary fuel engines into plasma reactors. Here, exhaust gases are recycled into a heated plasma that can be burned again. Even alternative fuels with high water content can be used, significantly increasing the efficiency of fuel engines and creating little to no polluting exhaust. By our definition, this may not be free energy – after all, fuel is still being consumed – but it is worth mentioning.

Examples of inventors in the field of plasma reactors include Philo T. Farnsworth[34] (also the inventor of television), Kenneth R. Shoulders,[35] Paul Pantone[36] and Dimitri Petronov.[37] Companies currently operating in this field include Aureon Energy[38] (formerly Safire Project[39]) and Brilliant Light Power.[40]

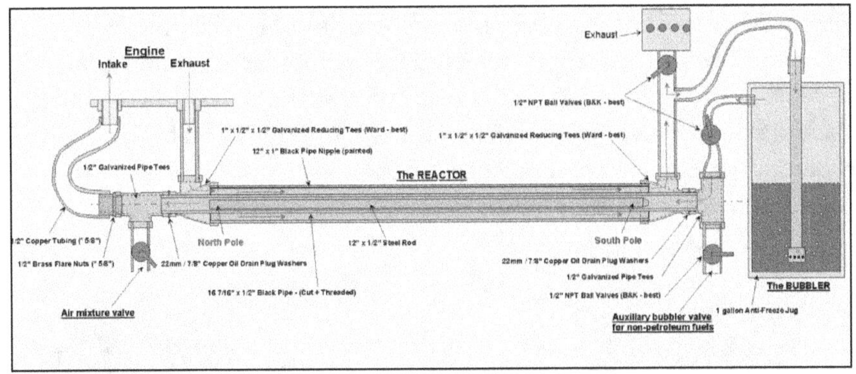

Figure 3.10: Schematic representation of Paul Pantone's Global Environmental Energy Technology (GEET). This device recycles exhaust gas by turning it into plasma that can be burned again in the engine.

LENR/cold nuclear fusion

Nuclear fusion has been studied for decades and is still seen as a viable solution to our shortage of clean energy. In a "practical" sense, however, hot nuclear fusion implies pouring billions of dollars into gigantic plants to maintain a fusion reaction at over a hundred million degrees Celsius. So, besides still needing a massive, centralized power generation infrastructure, this technology has not proven practical in reality. Nevertheless, it is researched anyway because nuclear fusion does not create unwanted radioactive waste, in contrast to the now mainstream process of nuclear fission.

In the late 1980s, the scientific community was considerably shaken by an experiment by Dr. Stanley Pons and Dr. Martin Fleischmann. They demonstrated that a fusion reaction was possible on a small scale and at room temperature. Their experiment showed excess heat in addition to several byproducts of a nuclear fusion reaction. The concept of *cold fusion* was born and caused a tremendous uproar because it seemed too good to be true. Scientists couldn't believe billions were already spent on achieving 'hot' nuclear fusion to no avail – and for no reason. However, replications of the experiment were not always found to live up to the same results – crucial for scientific acceptance – which increased distrust of the phenomenon. Articles of replications that succeeded were consistently rejected by established scientific journals because no scientific theory could explain the effect. As a result, the sincere scientists working on this new cold form of fusion began organizing their own conferences because they were dismissed as pseudoscientists and con artists at established scientific conferences.

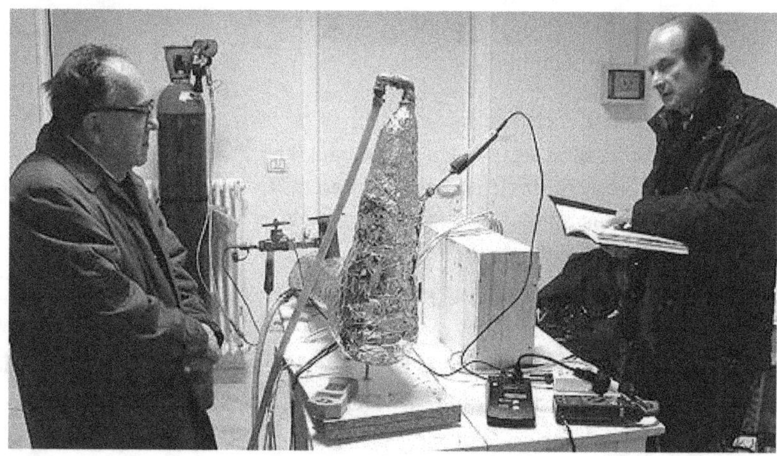

Figure 3.11: Sergio Focardi (left) and Andrea Rossi (right) with one of the earlier versions of the Energy Catalyzer (E-Cat). Since 2021, the E-Cat can be pre-ordered without obligation: see the website in the reference to Rossi below.

To this day, scientists are working in this exciting field. However, the question of whether this is genuinely free energy – the tapping of an inexhaustible and clean energy source – is beyond the scope of this book.

In the meantime, "cold fusion" has gotten such a bad name in science that people have started calling it Low Energy Nuclear Reactions (LENR). Eugene Mallove extensively researched cold fusion following Pons and Fleischmann's experiment for his book Fire from Ice.[41] Another example of cold fusion technology is Andrea Rossi's Energy Catalyzer,[42] described in detail in Mats Lewan's book An Impossible Invention,[43] and the company Brillouin Energy[44] is also active in this field.

Water technology

Water is truly a magical substance. Although water is abundant on our planet and most life couldn't live without it, it is one of the least understood substances. The small and humble water molecule behaves in ways we have yet to learn much from. Yet, so much fascinating research has already been done on water that you could devote a lifetime to it. Water purifies, heals, carries information and replenishes life. There are good reasons that the human body consists primarily of water.

In the field of free energy, water has taken a prominent role. Indeed, it is by itself a fantastic source of energy. What if you could pour tap water into your car's gas tank and then drive for miles without refueling? This turns out not to be just a pipe dream. Water (H_2O) consists of hydrogen and oxygen, which form a highly explosive mixture when mixed in their gas state. The trick is to split the water molecule into these two constituents efficiently.

Water splitting is possible in several ways. The most common method is *electrolysis*, in which water is exposed to a strong direct current. This pulls oxygen to the positive pole and hydrogen to the negative until the water molecule breaks, and the elements pair to form the gases oxygen (O_2) and hydrogen (H_2). The disadvantage of this method is that it takes a lot of electrical energy, of which much is wasted as heat.

A movement toward "sustainable hydrogen technology" currently exists in which hydrogen is first produced on a large scale through electrolysis or hydrocarbon combustion. Hydrogen is then transported in special highly pressurized tanks and converted back into electricity in vehicles in a so-called fuel cell before the energy is finally used in an electric motor. All these energy conversions and specialized transports are remarkably inefficient and, therefore, far from sustainable. Moreover, this idea again creates a dependency on a centralized hydrogen producer and hydrogen gas infrastructure, involving some severe and expensive safety measures.

It is much easier to turn water into a combustible gas mixture locally using resonance. By resonating an electromagnetic field in a specific way with the water molecule, it takes very little energy to split it.[45] More importantly, combustion of the gas mixture releases more energy than is needed to split the water molecule. Thus, the process is in a state of over-unity. In this way, many people have managed to convert their vehicles so that the only fuel needed is tap water. A reactor turns water into a combustible gas mixture that is then burned directly in the engine. The product of combustion is simply water vapor.

Examples of inventors of such a method include Daniel Dingel (see Chapter 4), Herman P. Anderson,[46] Stanley Meyer,[47] Ryushin Omasa,[48] Dennis Klein[49] and the Japanese company Genepax.[50]
However, the interest in water as a power source goes beyond fuel production. Viktor Schauberger[51] from Austria already observed at the start of the last century how water flows most naturally, thereby providing all of nature with life force. He developed technologies to revitalize water through vortices and meandering movements, giving it healing properties for plants, animals and humans. He also worked on an implosion generator that uses inwardly swirling water to generate energy. Even earlier was John E. W. Keely,[52] who built an "etheric generator" based on water vapor in the 19th century.

Figure 3.12: Stanley Meyer's famous dune buggy, which uses only water as fuel. There is a great deal of information on the Internet about his work. Unfortunately, Meyer died in 1998 at a dinner with potential investors due to poisoning.

In addition, there is the phenomenon of cavitation, in which tiny air bubbles in water implode to release a large amount of energy. The German company Rosch[53] has built a generator presumably using this principle, the *Kinetic Power Plant*, which is currently being sold under license to companies and municipalities in the Netherlands.[54] The list goes on, and there is so much to discover that we will not go into further detail. The curious reader should refer to the above sources and the sources at the end of this book.

Gravity technology

Gravity keeps us with both feet on the ground, but it turns out to be a source of energy of its own, after all. For centuries, many inventors have been trying to build a so-called *perpetual motion machine* with gravity as the driving force (see also the box on page 55). Most attempts were in vain because the energy it costs to lift something always equals the energy gained when you drop it again. Yet inventions have been made that seem to defy this law of conservation of energy.

Figure 3.13: William Skinner demonstrates his fully mechanical "amplifier" of gravity in 1939.[56]

For example, in 1939, William Skinner[55] created a device in which an assembly of weights circled each other in constant imbalance so that gravity always tended to drive the rotation. He reportedly achieved a COP as high as 120, with only a small motor to drive all the heavy machinery in his workshop. Gravity-based free energy research did lose ground over time, and very little can be found about it today.

How do we know if a free energy technology is genuine?

Research and development

As you can see, there is a large variety of ways free energy technologies may operate. The internet is full of alleged free energy devices; there is much more than we can mention here. However, we can safely assume that if something seems too good to be true, it usually is. Here, we are focusing on the exceptions that are at least worth further research.

So, how do we separate the wheat from the chaff? And once we have found a device that seems to work, how can we prove it does? From a computer screen, it is virtually impossible to know for sure. However, by doing research, you will naturally develop the intuition of whether you are looking at a free energy technology worthy of further investigation. Then, you will have to buy the device if it is available or reproduce it yourself to find out if it works. Neither option, however, is yet straightforward.

Very few free energy technologies are offered for sale. Those that are usually are so on a preliminary basis. In other words, the company only starts production once enough pre-orders are in. Sometimes, one has to buy a license to produce the device. This can be quite expensive, and often, the only guarantee that it works comes from (the company of) the inventor. However, public vetting contests such as *The Orion Project*[57] and the *Breakthrough Energy Project*[58] could further encourage the development of free energy technology.

Eventually, the technology may end up on the market, but this is yet another phase where free energy encounters much resistance. A lack of scientific understanding and verification by a reputable institute or a lack of mutual trust between investors, buyers and sellers are common obstacles. Moreover, there are still the ruling energy industries that have every interest in actively opposing such developments. After all, the energy trade is a huge revenue model worth trillions of dollars. We will see examples of these obstacles in the next chapter.

Building your own free energy device seems to be the better option. However, there is no definitive manual or theory of operation for any method of generating free energy. You will undoubtedly come across construction drawings and instructions when consulting the above sources. Still, they will not look like straightforward IKEA manuals, the result of which is known in advance. Free energy is a field of research and development – but don't let that discourage you. The advice for the creative and technologically savvy is to connect with like-minded people and study and reproduce an existing technology or support those working on one. See, for example, the list of websites and internet forums at the end of this book, where people from all over the world come together to further this research. And once you get started, what should you pay attention to when investigating free energy?

Criteria for a genuine free energy technology

A video of alleged free energy technology on the Internet says more than a thousand words, but more is needed to judge the technology for authenticity. The following are some essential criteria that a free energy device should meet. Of course, with the information we have, not every example mentioned before will directly meet all of these criteria, but because of the quality and quantity of information available, each of them is still worthy of further investigation.

Over-unity

The main principle of a free energy technology to demonstrate is over-unity: the output of useful energy must exceed the energy input. In this comparison, we must consider that those two energies – output and input – can be of different kinds. Let's take the example of a solid-state free energy technology, where we may compare high-frequency alternating current (output) with direct current from a battery (input). In the case of a plasma reactor, we may compare heat (output) with electricity (input). It's not impossible to work out these energy balances, but it requires at least a proper experimental setup and good math. Over-unity becomes much more evident if the device is self-powering and requires no external energy source. It then uses some of its generated energy to keep running; the rest is net free energy to be used at will. Such a system is the "holy grail" of free energy and can, in principle, be used directly as an independent generator.

No hidden conventional energy source

Of course, for this criterium to hold, there must not be a hidden battery in the device that provides the presumed 'over-unity' of energy. This is why transparency is crucial. A free energy device must be designed and demonstrated so that a hidden conventional energy source cannot be at work. By completely dismantling the device, we can see if it contains hidden batteries or fuels.

Endurance testing

Still, other energy sources must also be excluded, such as the wind, hidden wires, electromagnetic coils, or antennae that are out of sight. Hence, ideally, in addition to inspecting the device for trickery by completely dismantling it, it is also run in isolation for extended periods in endurance tests. After all, no small batteries or accumulators exist that will last that long and can be easily hidden. Such an endurance test, preferably conducted in different (remote) locations, demonstrates that a free energy device continuously taps into a completely independent energy source that does not run out. These criteria are best verified by independent, trained observers.

Independent reproduction

Finally, the device must be replicated by a third party. Once independent parties reproduce a free energy technology that also meets the above criteria, it is demonstrated that we are not dealing with a "lucky coincidence" but that a natural, universal principle is being employed.

Then, the technology can be confidently disseminated, replicated, and improved.

1 Nikola Tesla, "Man's Greatest Achievement," New York American, 1930 (written 1907).
2 Lipton, Bruce H. The Biology of Belief. Carlsbad, CA: Hay House, 2016.
3 wilhelmreichmuseum.org/
4 www.instructables.com/Pyramid-energy-does-it-exsist-Research/
5 vitalfluid.nl/water-purification-at-the-source
6 www.researchgate.net/publication/271616799_Gravity_and_Zero_Point_Energy
7 www.nytimes.com/2020/05/14/us/politics/navy-ufo-reports.html
8 www.the-sun.com/news/5053632/us-government-releases-1500-pages-secret-documents-ufo-programme/
9 coenvermeeren.nl/
10 www.youtube.com/c/DrStevenGreer55
11 www.rexresearch.com/gravitor/gravitor.htm
12 www.researchgate.net/scientific-contributions/Evgeny-Podkletnov-77360399
13 www.rexresearch.com/grebenn/grebenn.htm
14 www.rexresearch.com/carr/1carr.htm
15 rexresearch.com/hubbard/hubbard.htm
16 rexresearch.com/sweet/1nothing.htm
17 rexresearch.com/books/EnergyVacuumBearden.pdf
18 rexresearch.com/kapanadze/kapanadze.htm
19 www.youtube.com/watch?v=viGjXV38qvY
20 jnaudin.free.fr/meg/megv21.htm
21 rexresearch.com/johnson/1johnson.htm
22 rexresearch.com/yildiz/yildiz.htm
23 infinitysav.com/
24 www.youtube.com/watch?v=ab-4YBeCq-s
25 www.youtube.com/watch?v=L_jtiS41rDw
26 rexresearch.com/hendershot/hendershot.htm
27 segmagnetics.com/professor-john-searl/
28 www.brucedepalma.com/
29 rexresearch.com/evgray/1gray.htm
30 rexresearch.com/testatik/testart.htm
31 rexresearch.com/newman/newman.htm
32 johnbedini.net/
33 rexresearch.com/tewari/tewari.htm
34 fusor.net/
35 rexresearch.com/ev/ev.htm
36 rexresearch.com/pantone/pantone.htm
37 rexresearch.com/articles2/thorium.htm
38 aureon.ca/
39 safireproject.com/
40 brilliantlightpower.com/

41 Mallove, Eugene J. Fire from Ice. Infinite Energy Press, 1999.
42 e-catworld.com/
43 Lewan, Mats. An Impossible Invention: The True Story of the Energy Source that Could Change the World. Self-published, 2014.
44 brillouinenergy.com/
45 www.sciencedirect.com/science/article/pii/S1875389211006079
46 waterpoweredcar.com/herman.html
47 rexresearch.com/meyerhy/meyerhy.htm
48 rexresearch.com/ohmasa/ohmasa.htm
49 rexresearch.com/klein/klein1.htm
50 www.genepax.com/
51 www.goodreads.com/author/show/171550.Viktor_Schauberger
52 rexresearch.com/keely/keely.html
53 rosch.ag/uk_firmengruppe.php
54 www.zilverstroom.com/kinetic-power-plant/
55 rexresearch.com/skinner/skinner.htm
56 www.youtube.com/watch?v=X_JuaccKTPs
57 www.theorionproject.org/en/
58 breakthroughenergyproject.org/

Chapter 4: Inventors highlighted

Now that we have painted a picture of the ether and how several forms of free energy technology might operate to tap into it, let us explore recent history to highlight some high-profile examples of free energy inventors. Nikola Tesla was undoubtedly not the only one, as dozens, perhaps hundreds of people have developed free energy technology in his footsteps or on their own paths. Even though they have not (yet) succeeded in providing – or were not allowed to provide – the world with abundant clean energy forever, we stand on the shoulders of these pioneers to make their vision a reality.

Thomas Henry Moray (1892 - 1974)

"The universe is singing, and this symphony or frequency is what keeps every part of the universe and every atom in its proper orbit."

Thomas Henry Moray,
The Sea of Energy in which the Earth Floats

Figure 4.1: Thomas Henry Moray.

Moray's challenge

Thomas Henry Moray was born in 1892 in the American city of Salt Lake City. As a boy of fifteen, he was already experimenting with electricity, his greatest fascination. Still in its early days, it was a golden age for electricity technology. Only a few decades earlier, the foundations of the science of electricity had been laid, and little had changed since then. Researchers such as Michael Faraday and James Clerk Maxwell were Moray's heroes. Of course, Moray was also a big fan of Nikola Tesla, who at the time was turning the entire energy industry upside down with his alternating-current inventions. At the time, Tesla had said that all of space was filled with energy, just waiting for us to tap into it. That was the perfect challenge for Moray. Imagine what we could do with that abundance of energy!

Moray studied electrical engineering in 1907 and then worked as an electrical engineer for various construction and telephony companies. His primary interest, however, always remained Tesla's concept of Radiant Energy. Like Tesla, he sought to capture cosmic radiation, the Radiant Energy that permeates the entire universe, to convert it into electrical energy. In 1920, Moray was the victim of an industrial accident while working with telephone circuits, in which his eyes were damaged. This would eventually prove to be a blessing in disguise. Since he could no longer work as an engineer with companies, he was given more time to devote to his research on Radiant Energy.

Moray's Radiant Energy receiver

Starting in the 1920s, Moray developed a receiver for cosmic Radiant Energy. He used rare earth metals such as germanium and bismuth for this receiver, uniquely assembled to form a semiconductor material. This type of material later became incredibly important to the electronic industry. Today, semiconductors are used in all phones, computers, LEDs and solar panels. This alone showed that Moray was ahead of his time. His semiconductor material worked like a radio receiver for cosmic rays, in which the radiation could be sent in one direction into an electrical circuit. It was, in other words, a rectifier for Radiant Energy. Moray called his invention the *Moray Valve:* a "valve" allowing Radiant Energy to flow into a circuit.

He coupled his Moray Valve to an electrical circuit consisting of capacitors, coils and vacuum tubes, which served to amplify the energy received from the cosmos to considerable power. The whole circuit was solid-state (it had no moving parts) and was about the size of a shoebox. Attached to this box were a ground connection and an antenna, without which the device did not work. Moray connected his Moray Valve to it and tuned the entire instrument by stroking a heavy magnet alongside it. The high-frequency energy thus generated had a voltage of as much as 250,000 volts, while the power was as much as 50 kilowatts. For comparison, that's more than enough to power ten households continuously. In addition, the current produced by the device had the curious property of not heating components and wires like conventional electricity. As a result, it was quickly dubbed "cold electricity."

Figure 4.2: T. H. Moray during one of his Radiant Energy receiver demonstrations.

Moray made demonstrations before numerous scientists and members of the press. None of them were ever able to prove that there was any deception. On the contrary, even scientists who were highly skeptical of Moray's invention and came to prove him wrong came back from his demonstrations entirely convinced, agreeing that Moray could indeed tap into a hitherto unknown energy source to generate enormous power.

Moray showed that he could simultaneously power a large panel of lamps and all kinds of other large power consumers, such as irons, for as long as he wanted. None of the scientists present could provide an explanation for this. With the right equipment already existing, for example, harvesting several milliwatts of power by capturing radiation from nearby radio sources would be possible. But even when Moray's device was miles away from every potential electromagnetic radiation source – such as radio and electricity towers – it delivered undiminished power. Extensive, signed accounts of these demonstrations can be found in the book *The Sea of Energy in Which the Earth Floats* by John E. Moray,[1] the son of T. H. Moray, who has followed in his father's footsteps and is only too eager to take this technology further.

Setbacks

Despite numerous demonstrations before reputable scientists, Moray failed, even after seven attempts, to apply for a patent for his invention. The reason for these rejections was that a physical explanation of the device's energy output didn't exist – arguably not a good reason to reject a working device.

In addition, Moray and his family were often threatened and even physically assaulted. In an interview with Jeane Manning,[2] Moray's son John states that the family home and the laboratory where Moray worked were regularly broken into. He and his family were even shot at several times, forcing Moray to buy two pistols and put bulletproof glass in his car to protect himself and his family.

Felix Frazer, working at the Rural Electrification Administration (REA), managed to gain Moray's trust despite all this and secured a position in his laboratory as a "consultant" and "bodyguard." For two months, he was able to study the Radiant Energy receiver thoroughly. He wrote reports on the device, which he then forwarded to his "superiors in Washington D.C." Then, one morning, Frazer smashed Moray's receiver to pieces with a hammer. The same Frazer was later allegedly involved in a suspected attempt to kidnap Moray. In the process, Moray was struck in the leg by a bullet and barely managed to shoot his way out of the situation. These events are also detailed in *The Sea of Energy in Which the Earth Floats*.

Moray's final period

These intimidations left Moray barely able to continue his work effectively. He had clearly become entangled in a dangerous web of vested interests and people with wrong intentions. Furthermore, he was unable to obtain the necessary funds to make the expensive semiconductor materials needed for a new receiver. Its construction has always remained his best-kept secret, which is also one of the reasons why many others haven't been able to reproduce the device. Moray's patent applications, which should have been neatly filed with the patent office, have never been seen again. The best source for further research on his free energy device is some of his memos and notes, found in John E. Moray's book on his father's work.

Daniel Dingel (1928 - 2010)

"I didn't work 14 years day and night to come up with something for rich businessmen. I was able to come up with this car because I have always wanted to make life better for the people, especially the poor. I don't want to see hungry people anymore. We've suffered enough."

Daniel Dingel

A noble Filipino inventor

Daniel Dingel was born in 1928 in La Union province in the Philippines. Due to the early death of both his parents, he grew up in an orphanage. However, with the help of foreign donors, he was able to pursue a technical education as a mechanical engineer. His great dream was to put his inventive talent at the service of humanity, especially the poor.

Dingel came into the spotlight of the national press in the early 1980s. The reason for this attention was that he stated that he had been developing a car that required only water as fuel since 1969. This news surfaced just after the 1973 oil crisis when oil-producing countries in the Middle East imposed an oil embargo against the West. As a result, the oil price had risen nearly 300% within a year. The urgency to look into alternatives to oil as a fuel was high.

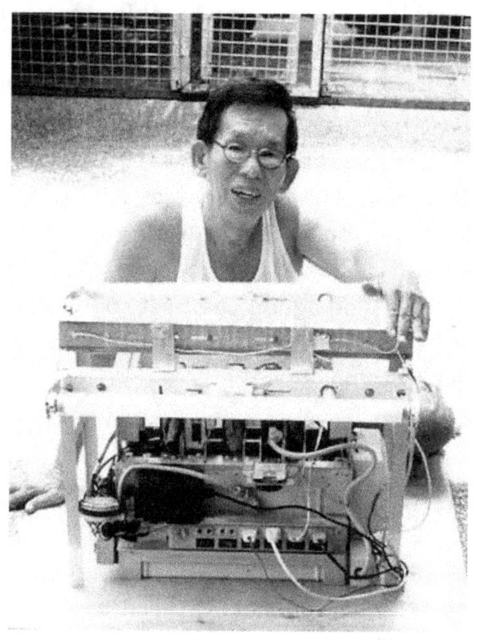

Figure 4.3: Daniel Dingel with his hydrogen reactor,
which uses only tap water as fuel.

A car that runs on water

Scientists from the Philippine government watched in amazement as Dingel demonstrated his invention during a series of test drives. For example, he drove 147 kilometers during one drive, in which the car had only consumed half a liter of gasoline and 15 liters of water. Dingel's car performed even better during another test drive in the United States. A car trip from Detroit to Florida (nearly 2,000 kilometers) was driven on two liters of gasoline and 60 liters of water.[3] This version of his car needed a small amount of gasoline to start the engine and to keep it idle.

Once the engine started moving the vehicle, the gasoline line was automatically shut off, and the engine ran on the combustible water-gas mixture.

But it was not only the Philippine government that was impressed. Several German, American and Japanese car manufacturers, the Philippine military and the Philippine Navy had also been convinced by Dingel's demonstrations and had become very interested in this technology.[4]

Figure 4.4: Underside of the hood of Dingel's Toyota Corolla, where he wrote down his 'secret.'

The chemistry of water as fuel

As noted in the previous chapter, producing hydrogen gas with the standard process of electrolysis is relatively inefficient. This is because using direct current for electrolysis consumes a lot of energy, most of which is lost as heat. Also, ordinary electrolysis usually requires a certain salt solution, and the yield of hydrogen gas is low.

In ordinary electrolysis, the water molecule (H_2O) is split into the ions – electrically charged atoms – of hydrogen (H^+) and oxygen (O^{2-}), which then immediately reassemble to form the gases hydrogen (H_2) and oxygen (O_2). We can then represent the reaction as a whole like this:

Figure 4.5: 'Standard' electrolysis of water, in which two water molecules are converted into two hydrogen gas molecules and one oxygen gas molecule.

Dingel's technology, however, worked markedly differently. He produced more gas and required much less energy to do so. At first glance, his technique was thus quite similar to that of a Bulgarian inventor named Yull Brown.[5] Brown was convinced that his electrolysis method formed not only hydrogen gas (H_2) and oxygen gas (O_2) but also ions of hydrogen (H^+) and oxygen (O^{2-}) in gaseous form. Consequently, this unique gas mixture has since been called "Brown's Gas" or "HHO" by those experimenting with it. Other inventors, such as Stanley Meyer and Ryushin Omasa, used various forms of resonance (electrical and mechanical) to create Brown's Gas. We could represent the resulting reaction like this:

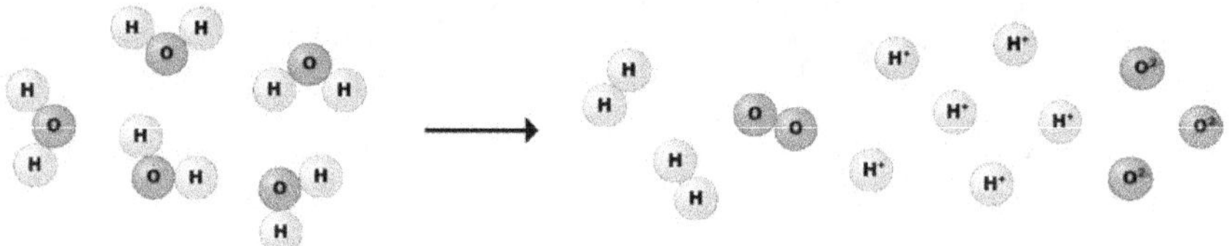

Figure 4.6: A suggestion for Brown's Gas production.
Here, five water molecules are converted into two hydrogen gas molecules, one oxygen gas molecule, six hydrogen ions and three oxygen ions.

Other ratios of molecules could also be possible in this reaction, but the essence is that more gas per water molecule is produced than in the standard reaction. Moreover, part of the resulting gas mixture is now electrically charged, thus becoming a plasma formed from water.

Brown's Gas has several remarkable properties. Under certain conditions, the gas can implode instead of explode when ignited, it releases more energy on combustion than ordinary hydrogen gas, and it can be used in a welding machine to melt metals that cannot be melted with standard welding equipment. It even has beneficial health effects on plants and humans.[6] According to several other inventors, an ordinary gasoline engine can run entirely on Brown's Gas with a few modifications (see the examples from the previous chapter) or at least be made more economical in combination with gasoline.

Lifelong struggle

The more often he gave demonstrations to companies and government agencies, the more disillusioned Dingel became. Officials were either eager to line their own pockets, or they congratulated Dingel elaborately on his fantastic invention, only to hear nothing more from them afterward. Offers from foreign companies were also highly disappointing. His dream of making the car cheaply available to ordinary Filipino citizens was of no interest to large companies. Dingel states the following about this:

> "They all wanted to make big money out of it, just that, at the expense of the common people. I am looking for selfless, honest investors. I haven't found any."

Two years before his death, Dingel was suddenly sued by a Taiwanese company with whom he had entered into a business agreement in 2000. He had not fulfilled (or rather, could not fulfill) his agreement to deliver prototypes of his car in the year 2000. He was sentenced to repay $380,000 and 20 years in prison; he failed to make good on both fines.

Dingel realized very well, in retrospect, that he may have made more enemies than friends with his invention.[7] He concluded that we humans were still too primitive in our interactions with each other. The bottom line was that humanity at large was not yet ready for his technology, which at that time could have provided an abundance of clean energy. Although Dingel is said to have converted more than a hundred cars in the Philippines to run on water, it is unknown where those cars are now – and the secret of his technology has never been made public.

Adam Trombly (1951 – present)

*"There is no such thing as free energy without enlightenment
and liberation. The technologies point to free energy.
We are the technology. We are the free energy."*

Adam Trombly

Figure 4.7: Adam Trombly, as seen in the movie Thrive.[8]

An unexpected introduction to free energy

Adam Trombly grew up in a family of scientists. His father was a biochemist, his mother a blood specialist, and his sister was on her way to becoming a biophysicist. The language of science was instilled in him at an early age. Trombly's father worked with the Central Intelligence Agency (CIA) on top-secret military projects but died at the age of 38. Trombly claims the CIA poisoned his father because he disagreed with its policies – an extraordinarily intriguing story worth reading.[9]

Years later, 15-year-old Trombly snooped around in the attic and found some of his father's secret notebooks. In them, his father described how he had encountered technologies of "extraterrestrial origin" during his research. Technologies that defied all known laws of nature and "exceeded his wildest imagination." They included saucer-shaped spacecraft and how these were powered with an unlimited, clean energy source. Apparently, technologies that could instantly solve all of humanity's so-called problems already existed in secret.

This was what Trombly needed to ignite his passion for free energy. He educated himself in many fields, including physics, electrodynamics, astrophysics and climate science. He also immersed himself in understanding humanity's current situation and began working on solutions.

Together with a friend, the famous architect and scientist R. Buckminster Fuller, he founded Project Earth in 1970 – a "human design experiment" to create a network of people who would share solutions for fully sustainable humanity and Earth. Buckminster Fuller warned Trombly against developing free energy technology, stating that if he succeeded, there would be hell to pay. Here, he was referring to the social, scientific and corporate resistance to free energy that Trombly would face. Few people were open to the idea, and many people in positions of power certainly were not.

The Closed Path Homopolar Generator

Nevertheless, Trombly got to work and, in 1980, developed the Closed Path Homopolar Generator with his colleague Joseph Kahn. This motor-generator taps into the zero-point energy field (the ether) to provide more energy than it needs to run. They even received an international patent describing the device in considerable detail.[10]

The operation of the generator is based on a nearly 200-year-old invention by Michael Faraday, one of the founders of electrical engineering. Faraday discovered that when a copper disc spins in a stationary magnetic field, a current begins to flow between the axis (the center) and the edge of the disc. This was, in fact, the first version of the dynamo – a generator that generates current using a magnetic field.

Trombly and Kahn designed their generator with the magnet attached to the disk and thus rotating with it. They also ensured that the magnetic field lines covered a complete, closed path through a thick steel casing – hence the term "closed path." In this way, there was nowhere for the magnetic field to 'leak' out of. Because of these features, combined with drawing energy from the ether, this device achieved an efficiency of 492% (a COP of 4.92).

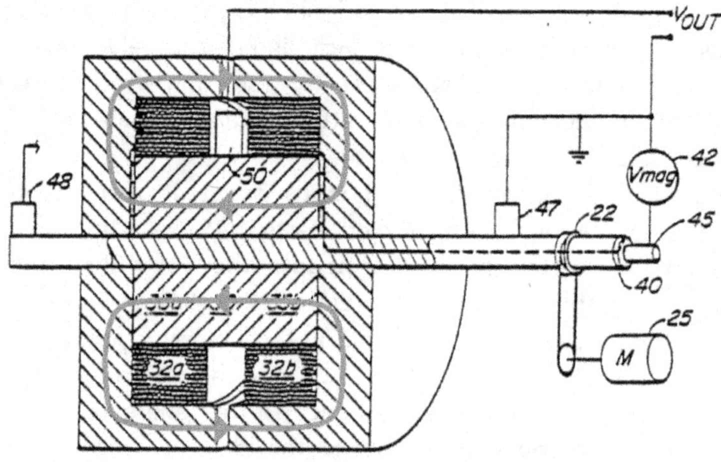

Figure 4.8: A cross-section of the rotor (the rotating part) of the Closed Path Homopolar Generator from Trombly and Kahn's patent.
In grey is the "closed path" of the magnetic field.

To prevent Trombly's invention from actually seeing the light of day, he was silenced by the U.S. Department of Defense in 1983. He was not allowed to further develop or disseminate information about the device, or he would be jailed for ten years. However, the international patent had already been registered, and around 1985, due to standard procedures, it automatically entered the public domain.

At the same time, another engineer was working on a similar homopolar generator. Inventor Bruce E. DePalma called it the N-machine.[11] Along with Trombly's patent, he forwarded the idea of the homopolar generator to a third engineer from India, Paramahamsa Tewari.[12] At the time, Tewari had been working on a new theory of the structure and energy of space for years and was looking for an experiment to test his theory. Trombly's device came right on cue, and Tewari immediately replicated it. With his own design, he achieved an efficiency of 264%, independently verifying the principle of Trombly's machine. Both DePalma and Tewari have extensively documented their work on the homopolar generator on their websites.[13] [14]

Inertia and resistance

After the publication of Trombly's patent, his silence was "officially" lifted. He continued his research in zero-point energy and, in the late 1980s, designed a solid-state free energy technology with another colleague, David Farnsworth. It was called the Piezo Ringing Resonance Generator and had a COP of a whopping 54 – the device generated as much as 54 times more electrical energy than it consumed.[15] This was such an important invention (the superlative of his previous device) that Trombly was allowed to present it at a United Nations meeting in New York in June 1989.

Initially, the demonstration for the entire UN company was to be held in one of the main auditoriums. At the last minute, however, for "security reasons," the demonstration had to be held in a smaller company and a different room from the main auditorium, separate from Trombly's speech. The demonstration was attended by about 50 people, including technical experts and businessmen from Wall Street. Some of them were moved to tears as they understood the implications of such technology. Trombly was then allowed to address the larger gathering at the UN and received a standing ovation afterward. A few days later, Trombly and Farnsworth also provided a demonstration before members of the U.S. Congress.

After all these demonstrations, one would expect everyone to be excited and motivated to advance Trombly's technology further. However, the opposite was true. Trombly and Farnsworth's lab was raided by US Marshalls (a federal police agency) in March 1994. All sorts of expensive equipment, including everything related to power generation, were loaded into trucks and taken away. From those within the UN and Congress who had become genuinely enthusiastic, the inventors heard nothing more.

In addition to these material setbacks, Trombly said he has also been harassed and physically assaulted. For example, he has been poisoned several times, with his wife even having to resuscitate him once.

The ongoing mission

Fortunately, Trombly is still alive at this writing and has not ceased voicing his truth. At the moment, his international advocacy is expanding our consciousness to embrace these technologies collectively. In an interview with Mark Schreibner in 1999, he said:

> "At this point, I am recommending that people pray like they never have before for Divine or at least Benign Intervention. It is unlikely that the boys in Washington and other world capitals are going to implement any world healing policies unless the populations of the world rise up and unequivocally demand change. We don't have much time."

Garret Moddel (1954 - present)

> "What has been the response of the scientific community? What are they saying, given that this potential for something really radical has been described? [...] None. [...] We've heard nothing from them."

Garret Moddel

Figure 4.9: Garret Moddel

A distinguished professor on the frontier of science

Garret Moddel was born in 1954 in Dublin, Ireland. He studied electrical engineering at Stanford University and received his master's degree and doctorate in applied physics from Harvard. After spending four years in research at a company specializing in solar cells, he became an assistant professor at the University of Boulder, Colorado, where he is a professor of electrical, computer and energy engineering to this day.[16]

About a decade ago, during a sabbatical, Moddel became interested in the field of parapsychology, which studies the interaction between physical reality and (human) consciousness.[17] Although this field is still in the taboo area of mainstream science, Moddel has already conducted some startling research in it. For example, in his attempt to use a group of 10 students to predict the rise or fall of the Dow Jones stock index seven times, the group succeeded every time.[18]

At the same time, Moddel has been publishing scientific papers since 2003 on his research into zero-point energy and how we can tap into it – a study that perhaps generates even more frowning eyebrows within the scientific community. According to Moddel, there are roughly two reasons for this. First, this new science is not yet well-grounded, making claims that are not (yet) scientifically correct. Second, there is unprecedented resistance within mainstream science because it clings convulsively to certain dogmas.[19]

One such dogma is that the entropy of the zero-point energy field is maximal. In other words, the energy in the zero-point energy field is maximally evenly distributed across space, and there is no way to "gather" that energy effectively. In other words, the previously explained notion of syntropy doesn't exist. However, no experimental evidence exists that supports this. The idea comes from the second law of thermodynamics, which implies that it is impossible to extract heat from a system in which the temperature is the same everywhere. This law worked perfectly to describe the invention of the steam engine at the end of the 19th century – a closed system that could produce no more energy than it consumed in terms of fuel. Since then, the law has been applied unaltered to all new fields of science, including quantum physics. But does that law also work for open systems? Does it also apply to the zero-point energy field or ether? Moddel does not think so. He has been publishing research on zero-point energy that we can safely call groundbreaking.

Zero-point energy from a Casimir cavity

In 2008, Moddel and his colleague Bernard Haisch published a patent called Quantum Vacuum Energy Extraction.[20] They describe a device that pumps gas around in a closed tube, forcing it through a grid of Casimir cavities. In Chapter 3, you already read about the Casimir effect, in which the space between two thin mirrors that are very close together contains fewer "vacuum fluctuations" (zero-point energy) than the space outside it. This creates a pressure that pushes the mirrors together.

A Casimir cavity is a hole in a material small enough to "strip" gas atoms of some of the energy continuously provided by the zero-point energy field. Once the gas atom enters the cavity, that energy is released as electromagnetic radiation that can be captured. When the gas atom leaves the cavity again, the gas atom's original energy is "replenished" by the zero-point energy field. Thus, each gas atom can be reused repeatedly to extract energy from the zero-point energy field. The device is currently being further developed by the company Jovion,[21] which is seeking funds to bring it to market.

Even more recent is Moddel's research from 2021, in which he uses a tiny device – many times thinner than a human hair – with a Casimir cavity to generate a continuous electric current by tapping into the zero-point energy field.[22] Admittedly, it's a very small current, but by linking these tiny solid-state devices, we soon get a considerable power output comparable to a solar panel. Moreover, this "solar panel" does not need sunlight, so it can even be stacked into a block while providing energy day and night.

In a fascinating presentation on his work, Moddel discusses the implications of this technology.[23] He mentions cell phones that never need recharging, electric cars, trains, ships and planes with their own energy supply and thus an infinite range and no more emissions from burning fossil fuels anywhere.

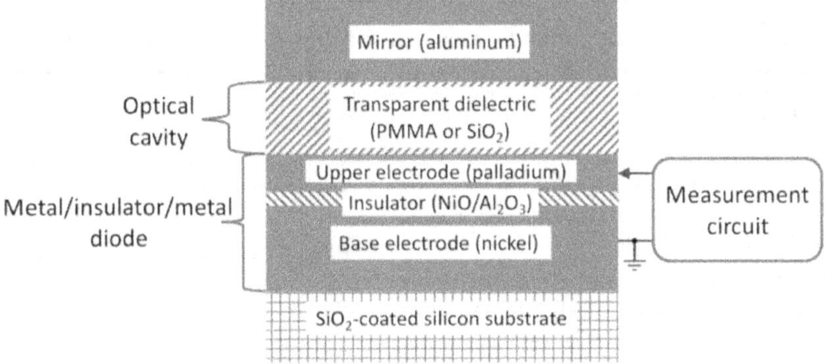

Figure 4.10: Schematic representation of the solid-state device from Moddel's research. The device uses a Casimir cavity ("optical cavity") to create a permanent imbalance in the zero-point energy field (the ether), causing an electric current to flow from the "base" to the "upper" electrode.

The response of established science

The research by Moddel and his group was published in two reputable science journals. Yet, as the quote at the beginning of this story indicated, any response from fellow scientists remains absent. A deafening silence persists. Meanwhile, Moddel also released other research that was much less groundbreaking and important, from which more than enough press attention and interviews emerged. Why doesn't this happen with his most important work to date? Moddel himself explains:

"I suppose people just don't know what to make of it.
They just don't believe it."

Despite this silence from the corner of science, Moddel applied for a patent for this technology in February 2022.[24] Time will tell if his developments will be accepted and can reach the market.

As is evident from the stories above and many others, there is more than just a technological challenge to overcome in developing free energy technology. Something essential must also change our attitudes toward major change, how we look at the world and our institutions. Moreover, how we look at ourselves as human beings is fundamental, and we will return to this later. First, let us look at the much-needed evolution of science.

1 www.naturalphilosophy.org/pdf/ebooks/Moray-TheSeaofEnergy.pdf
2 Manning, Jeane and Garbon, Joel. Free Energy. Breda: Obelisk Books, 2019: p. 110.
3 fuel-efficient-vehicles.org/energy-news/images/pages/Daniel-Dingel-Manila-Times-94.jpg
4 fuel-efficient-vehicles.org/energy-news/?page_id=928
5 www.rexresearch.com/hyfuel/ybrown/4014777.htm
6 eagle-research.com/what-is-browns-gas/?doing_wp_cron=1663166505.4471309185028076171875
7 www.youtube.com/watch?v=pywNSMRiFJQ
8 www.thrivemovement.com/the_movie
9 www.rexresearch.com/trombly/trombly.htm
10 Trombly, A. D. & Kahn, J. M. (1982). Closed Path Homopolar Machine (Int. Patent No. WO82/02126).
11 rexresearch.com/depalma/depalma.htm
12 rexresearch.com/tewari/tewari.htm
13 www.brucedepalma.com/
14 www.tewari.org/
15 projectearth.com/about/adam-trombly
16 ieeexplore.ieee.org/author/37315946500
17 skeptiko.com/garret-moddel-brings-psi-to-colorado/
18 www.colorado.edu/faculty/moddel/sites/default/files/attached-files/smith_sse10_abstract.pdf
19 www.youtube.com/watch?v=1FPf3PiqiZY&t
20 Haisch, B. & Moddel, G. (2008). Quantum Vacuum Energy Extraction (U.S. Patent No. 7,379,286).
21 www.jovion.com/home
22 www.mdpi.com/2073-8994/13/3/517
23 www.youtube.com/watch?v=2tGRhTXKh8A
24 Moddel, G. (2022). Quantum Noise Power Devices (U.S. Patent No. 11,258,379).

Chapter 5: Evolution of science

"The day science begins to study non-physical phenomena,
it will make more progress in one decade
than in all the previous centuries of its existence."

Nikola Tesla

Free energy technology brings such radical change that it requires an entirely new science. This is easy to say, but such a giant leap does not come without a struggle. Technology and science have been closely linked from the very beginning. Together, they have a massive impact on how we live – we currently see this as never before. This chapter explores the meaning of technology and science and how they can evolve in transitioning to a world where free energy flourishes. Tesla knew long ago that we must look beyond physical reality and include the spiritual dimension – the dimension of consciousness – to understand the world and our evolution within it.

What is technology anyway?

"There is a difference between progress and technology. Progress
benefits mankind. Technology does not necessarily do that."

Nikola Tesla

Technology is increasingly seen as the measure of human progress. In the modern era, "innovation" (in technology) has become the magic word with which many institutions promise to contribute to improving the quality of life for everything and everyone. But when does technology *benefit* the world? And who decides?

The term "technology" comes from Greek and loosely translates to "the study of skill." Technology is a means to make achieving a goal easier, such as making a cup of coffee with a coffee machine or plowing a field with a plow and a tractor. At first glance, technology seems to enrich our lives by giving us more freedom in what we can do. Technology can take work that we find burdensome or uninspiring out of our hands, allowing us to focus on more enjoyable and meaningful things. In short, technology indeed aims to improve our quality of life.

However, if we look at our current technology, it is beginning to take on a life of its own – literally. It seems as if the addiction to smartphones and the arrival of an all-dominant artificial intelligence are inevitable. Technology is, as it were, taking us for a ride. It often turns out that if we introduce one technology to solve a problem (whether or not it's a *real* problem), several new issues arise, for which we then have to invent yet other technologies.

This may provide exciting new business cases for tech entrepreneurs, but it is an undesirable, runaway situation. One could even say that we are moving backward in some fields of society, considering the damage we have caused with technology over the past hundred years.

On a deeper spiritual level – or level of consciousness – our entire concrete living world reflects our inner values and ideals. The freedom to do more and having to do less, which we now try to acquire with technology, is not the freedom that truly matters. True freedom can only be found within our own being. If we feel addicted and subservient to technology, we have indeed lost touch with that innate sense of freedom.

However, if we restore that connection, our outer world and technology will begin to reflect this. Hence, human progress, science and the development and use of new technologies are at a critical crossroads. Innovating for the sake of innovation no longer makes sense. From now on, we must focus on our inner state of freedom to let our outer reality correspond to it. Simply put, technology alone will not make the world a better or happier place. Instead, we need a more holistic view of the world, where spirituality and technology go hand in hand. This is essential in developing a radically new energy technology like free energy, which we discuss further in the final chapters.

Today's academia

> *"All truth passes through three stages. First, it is ridiculed.*
> *Second, it is violently opposed.*
> *Third, it is accepted as being self-evident."*
>
> Arthur Schopenhauer

Science and technology are driven primarily by those who have studied in these fields: academics. Universities and laboratories, where most research occurs, are closely intertwined with other branches of society, such as big international business and politics. Even though we want science to be pure and independent, in reality, it often is not. Universities are sponsored by large corporations, which consider an influx of intelligent people and the research they do as a revenue model. Through business and lobbying, the government makes and changes policies regarding the conduct of science. This directly or indirectly determines which research is allowed and thus receives funding.

For instance, a real breakthrough in energy technology at a university is unlikely to happen because large companies in the energy sector would see their business model vanish into thin air. Not only would it disrupt the status quo of power and control, but it would make the public independent from institutional power altogether.

In this way, research is only being done on that one extra percent efficiency of a solar panel or combustion engine. These are the technologies that eventually reach the market.

The same could be said, unfortunately, of the medical and pharmaceutical industries. These industries are not fully geared toward solving rampant health issues definitively because they would lose their reason to exist if they were. For the same reason, they mainly focus on the symptoms of health issues instead of the (often straightforward) root causes, such as diet and lifestyle.

In addition, many scientists themselves have difficulty accepting significant change. The notion that we can harvest unlimited amounts of energy from the ether is unacceptable to many because, if that were true, their academic world would be largely invalidated. You can't just tell a nuclear physicist specializing in atomic fission that we have been able to harvest energy from anywhere in the universe for over a century, which would imply that his life's work has been futile.

Even some physical laws and principles he grew up with would have to be cast away or at least heavily revised. That is why most scientists continue along the same path, not least for fear of being ostracized by their colleagues or because of losing their sponsorship and subsidies.

Again, free energy concerns major industrial interests more than any other disruptive technology.

Many scientists also feel that they are the guardians of the established order and that the answers to the essential questions have already been given. These answers, they believe, have been provided by the great minds of the Earth, such as Einstein and Newton. This, too, ensures that radically new ideas are hardly explored, let alone accepted. Such conditions in academia continually put the handbrake on actual progress. New generations of scientists in universities are trained in this climate. Due to its dogmatic adherence to so-called established laws, science has increasingly taken on the traits of a belief system, and according to some, it has already become a religion.

We cannot blame the man or woman in the scientist. The vicious circle in which science currently finds itself is perpetuated by a mindset of lack and fear that we all have to some degree. This is another argument for concluding that the evolution of science must ultimately go hand in hand with an inner, spiritual evolution.

The merging of spirituality and science

Science searches for truth and this quest for truth is inherent in the curiosity of humankind. What is real? What exists and what doesn't, and why? What is existence anyway? These questions have been asked since the beginning of humanity (and perhaps even earlier) and are the origin of spirituality and science. Where the two branches have gone hand in hand throughout history to a large degree, the Enlightenment (ironically) caused them to be driven further and further apart, to the point where it now seems they have nothing to do with each other. Yet, they still share the same goal of experiencing truth. As we find out that science as we know it has reached a crossroads and falls back on the essential spiritual questions as stated above, it becomes clear that the two have never been separate.

This crossroads is signaled by the advent of quantum physics and its related experiments. From the seventeenth century until the start of the twentieth, science had been limited to the Newtonian interpretation of reality, which assumed that a purely materialistic world existed "out there" within which consciousness arose by chance. That particular phenomenon – consciousness – was reduced to nothing more than a random electrochemical process in our brains. Gradually, this view is changing to a picture in which consciousness creates the stage on which the universe plays out all its roles.

In the scientific method, we successively observe the reality outside of us, predict the natural principle behind it, experiment to test that prediction, and finally, draw a conclusion to see if our prediction was correct. In each of these steps, we are forced to make certain assumptions. However, there is one fundamental assumption we can never test with this method – the assumption that there is a "world out there" to begin with, an objective world independent of our subjective perception. Certain experiments in quantum physics clearly demonstrate that this assumption is unfounded: consciousness can and will influence reality at every step of the way. This points to what mystics have known for thousands of years: consciousness and the universe – the observer and the observed – are two sides of the same coin and, therefore, inseparable. The universe *is* consciousness.

We will not dwell further on the wondrous discoveries of quantum physics, as numerous works elaborate on them, such as Lynne McTaggart's *The Field*[1] and Michael Talbot's books.[2][3] The bottom line is that there is no experience of the world without our awareness of it. Everything is interconnected, and a single field of consciousness is somehow at the core of it. This is the essence of spirituality: the direct knowing and experience that everything exists and comes from one spirit, one universal consciousness.

1 McTaggart, Lynne. The Field. New York: Harper Perennial, 2008.

2 Talbot, Michael. The Holographic Universe. New York: Harper Perennial, 1991.

3 Talbot, Michael. Mysticism and the New Physics. New York: Penguin Books, 199

Chapter 6: Implications of free energy

The English language has a very fitting expression for technologies like free energy: *disruptive technology*. In short, free energy turns our world upside down completely. In the first chapter, we briefly discussed the importance of free energy and mentioned several ecological and sociological consequences. In this chapter, we look at these in more detail. What should we consider if we want to introduce free energy smoothly? What happens when energy becomes cheap, clean and available everywhere in the short and long term? Finally, we will highlight why free energy can only flourish in a society hardly comparable to the one we live in today: a society of true peace.

Free energy, now!

The current state of global society

We can look at the current state of the world in different ways. From an optimistic perspective, humanity has never been more prosperous. More people have access to clean water[1] and electricity[2] than ever, both in percentage and absolute terms. Although the world's population has grown significantly in recent decades, population growth is slowing down. We may reach a point where population growth is net zero this century.[3] In addition, sustainable initiatives are steadily gaining ground in most countries worldwide.[4]

From a more "activist" perspective, the development of society – for example, in the areas of climate and the economy – is not going fast enough or even backward. Still, two-thirds of the world lives in poverty (with an income of less than ten US dollars a day),[5] the eight wealthiest people own as much capital as the poorer half of the world's population[6] and the gap between rich and poor is only widening.[7]

Sustainable alternatives in energy generation, production and recycling that have existed for over a century still cannot compete with the power of big business. At an accelerated rate, the Earth is being depleted. Resources are running out, and natural areas are being robbed and polluted to feed a growing consumer society.

Free energy is a game-changer: truly sustainable and limitless energy could become available to all humanity.

Rapid introduction of free energy

What happens if we introduce a technology tomorrow that can continuously generate clean energy on a large scale? Let's conduct a thought experiment.

Suppose we have a solid-state free energy technology with an adjustable output of up to 50 kilowatts. The cost, energy and raw materials to produce the device on a large scale are covered. Therefore, the device can be quickly deployed in many places. Individuals can share one device, providing continuous power to ten average households. We can install one device in an electric car to give it a virtually endless range – without batteries. Other forms of transportation, such as public transport and goods transport, can be supplied with a stack of free energy units linked together, and so can factories, farms, greenhouses and other large energy consumers. This may sound impossible now, but let us follow this line of thought a little further.

Most prices of goods and services will drop dramatically since a product's energy and transportation components make up most of its price. Once the production of free energy equipment takes off, all energy consumers rapidly become independent of a central energy supplier. Suddenly, much less energy is needed from fossil fuels or the sun, wind and water. The demand for centralized energy will decrease sharply. Since the energy required can be generated anywhere and anytime on demand, centralized energy suppliers are no longer needed. The problem of energy storage, therefore, is also immediately eliminated.

As a result, wind and solar farms may be shut down because they suddenly cost more than they raise in profit. Demand for oil and gas will be drastically reduced. The current energy market will shake to its core: within weeks, energy will become so cheap that it will be difficult to attach financial value to it. The US dollar, based primarily on oil trade, will collapse – although this is already underway, partly because of the enormous US debt and inflationary action. Other currencies, such as the euro, will undergo the same devaluation. It will create a kind of chaos we have never seen before – a chaos of simultaneous abundance and collapse. Globally, hundreds of thousands of jobs in the energy, financial, and other sectors will soon become redundant. Such a scenario has sociological consequences that cannot be overseen at a glance.

So far, the introduction of free energy sounds not all rosy but also quite turbulent. This is what disruptive technology means. This little thought experiment serves to burst the bubble of the illusion that free energy instantly lands us in a utopian world.

Centralization

The previous train of thought also shows how critical our energy economy is and how it is intertwined with every facet of our daily existence. There are numerous short-term consequences of rapidly adopting a free energy technology that we have not mentioned. We have all invested so much in our current energy economy that everything depends on it. If one card is removed from this house of cards, the whole thing comes crashing down. Now, you likely understand better why resistance to free energy is so great.

How did we get here? Over the centuries, human society has generally become too centralized. This is firstly because we began to equate centralization with efficiency. Of course, in an ideal world, it is easier to control a system when it is maximally centralized. In the best interest of the citizen, changes can be made relatively quickly and uniformly, and ideally, this saves costs.

Second, in a centralized system, much of society is "unburdened" because relatively few people bear responsibility for it. A great example of this is the development of the supermarket. After the Second World War, more and more small grocers and butchers were absorbed into or competed away by supermarkets. Where the responsibility for food supply was previously spread over many local parties, most people (in Western countries) now depend on one of the larger supermarket chains.

Society has effectively traded some of its responsibility – and, with that, freedom – for convenience.

This leads us to the two main drivers of centralization that go hand in hand: power and control. This is often symbolized by the 1% controlling the 99%. If power and control are the ends, then profit and scaling up the economy are the means to those ends. To quote Henry Kissinger, former Secretary of State of the United States:

"Who controls the food supply controls the people; who controls the energy can control whole continents; who controls money can control the world."

By this, Kissinger meant that if one centralizes and controls the money system, one inherently gains power over the energy and food supply. Businesses in these three sectors – finance, energy and food – are owned globally by only a handful of international stakeholders. Small, local companies are quickly competed away, and citizens become utterly dependent on these central systems by going along with this (by choice or not).

Centralization has made society's "ecosystem" very vulnerable. If, for instance, the computer systems of a major energy supplier are hacked, half a country could suddenly be without power. And if the European Central Bank gets into trouble, European savers can no longer access their money. After all, all national banks are linked to the central bank.

It becomes clear that such a climate is unfavorable for introducing free energy. Free energy, by definition, implies decentralized energy management. As a result, private individuals and (small) companies can meet their energy needs locally. That's why it is time to ride the already ongoing wave of healthy decentralization in our society, creating a new economy from the bottom up.

A new economy

Decentralization

As outlined above, chaos breaks out if we swiftly introduce free energy on a large scale. Our current society is simply too centralized. The challenge, therefore, is to work together to find a more decentralized and flexible way to introduce free energy into society. We may introduce free energy in self-sufficient communities large enough to provide all basic needs such as clean water, food, shelter, energy and healthcare for an entire group of people – and sufficiently decentralized to be independent of other self-contained communities and a central government. If the economy is also local, it is easier to shape, and changes impact other local or regional economies less. In the Netherlands, there is already a movement to make rural regions self-sufficient in this way. Society 4.0 by Dr. Bob de Wit, a Dutch professor of strategic leadership at Nyenrode University, is an example.[8]

Local economy

A local economy can thrive when there is a natural and balanced exchange of goods and services in a community, where the value of money represents *real value* – we will expand on this later. This used to be the case until banks came into existence that began to extend credit and make a profit on it by charging interest. As a result, credit was no longer based on 'real value' like gold but was issued on paper: money was essentially created out of nothing. Thus, the devaluation of money – inflation – was built into the system of central banks.

Unfortunately, the dollar, the euro and many other currencies are already at an advanced stage of devaluation. That means that the money people have earned and saved through their labor and creativity is becoming worth less and less. The Weimar Republic in the 1920s shows how that can end: a loaf of bread that cost 4 Marks in December 1921 could be bought for over 200 billion Marks two years later.[9]

Now, the accounts of banks and governments – both in the US and the EU – can no longer be settled because the accumulated debts (through loans with interest) have become too large. Moreover, with their endless printing of euros and dollars, central banks do not provide a sound foundation for a healthy (local) economy.

As a side note, central banks have been developing the "solution" to this failing financial system for several years. Namely, they want to centralize the current money system even further through *Central Bank Digital Currency* (CBDC): a fully digital replacement of every major currency now in circulation, to be issued and managed by a central bank (such as the European Central Bank)[10] through a government. In the meantime, cash should be slowly but surely phased out.[11] In practice, this means that the government – and, by extension, the central bank – will have insight into every transaction and can even determine how much we can spend on what and how much we can save.[12][13] The government can set limits on certain transactions or travel, for example, under the guise of a personal CO_2 credit,[14] and enforce them strictly through a CBDC. Your debit card simply declines to make a transaction once you reach your limit. A dystopian situation, where governments and banks can exercise even more power and control – something that can hardly be seen as a solution to our failing financial system for ordinary citizens.

For these reasons, starting with a clean slate and allowing a local economy to emerge with its own currency is more interesting. There are already examples of new local currencies flourishing in the Netherlands, such as the Florijn[15] and the URA.[16] Digital crypto-currencies such as Bitcoin are not tied to any particular region or state and thus may also play a critical role in decentralizing the financial system.

Back to free energy. If we look at how to smoothly integrate free energy, introducing it in a local, self-sufficient community is a lot more manageable than doing it right away nationwide or globally. If such a community already generates energy by its own means (e.g., solar and wind), we can easily replace those systems with free energy technology.

A self-sufficient community in today's era pays for its energy only for the initial cost and maintenance of, say, wind turbines or solar panels. A central energy supplier is no longer involved: the community has its own energy network, or each house may even have its own energy source. The value of the local currency is not directly linked to the energy supply, unlike the US dollar, which is directly related to the price of oil.

Most of the economic value of a self-sufficient community is represented by its local services and goods. That situation hardly changes with the introduction of free energy in the community. However, it would eliminate the disadvantages of solar panels and wind turbines, such as fluctuating power and low efficiency. Energy suddenly becomes much cheaper, but the local economy experiences a lesser shock than our centralized economy would. A baker still bakes his bread, and a greengrocer grows his vegetables at less cost, so products become cheaper locally.

This example shows what free energy can do to the economy if introduced on a limited scale. Decentralizing energy generation and economy at the proper scale makes the transition to free energy viable. If there are more such communities, they will be eager to follow the example of those who have integrated free energy. Free energy will then take off in different places. The traditional energy industry in the still largely centralized society thus gradually loses influence. Gradual phasing out is possible as more and more self-sufficient communities emerge until the moment when no one benefits from centralized energy generation anymore.

Now, let's go a step further and look closer at the purpose of the (local) economy to get a better sense of the potential impact of free energy. If the baker bakes his bread because he enjoys it, someone else may brick his house because that is his talent and passion. A plumber then repairs the bricklayer's pipes. Just as the baker bakes bread for everyone in the community, the greengrocer grows his vegetables for everyone and gets what he needs to do so. In a local economy, it becomes easier to see that we want to fulfill each other's wants and needs and that money is just an accounting tool. Money is a unit of measurement for value, but it hardly holds any real value and should not be an end. Instead, it is a means to reflect equivalence in an interaction of giving and receiving.

One might wonder whether we ultimately need money to give or receive, much less whether we should accumulate money in heaps to consider ourselves "rich." Thus, a new (but actually ancient) concept of economics, the *gift economy*,[17] is that everyone gives to society from their passion and receives what they need. Perhaps that is too rosy a picture of the future, but it is a vision we could aim for. In any case, the local community with its own economy shows us what should be our priority: meeting our basic needs fairly and sustainably.

What is value?

Because introducing free energy requires us to rethink our economy completely, it also gets to the heart of what we humans mean by *value*. In an absolute sense, one could argue that everything in the universe is of equal value or that nothing of itself has value. Both views are equally valid because we have no frame of reference in an absolute sense. That means we cannot define such a thing as *intrinsic* or *objective value*. It always remains a value that we ourselves have assigned.

We must realize that we are always looking through human glasses, and only through these can we determine what is valuable to us. Value is often based on feeling and, therefore, subjective.

At the individual level, for instance, an antique clock belonging to Grandpa is priceless to some but worth no more than a hundred euros to others. At the collective level, value is mainly determined by supply and demand and the statistics of large numbers. If a product becomes scarce while its demand remains high, it becomes worth more. Some goods are kept scarce on purpose to stabilize or increase their value. We use money to come to some consensus on the value of a product or service, but this is arbitrary and a matter of convention. Nowadays, we even let algorithms determine the value of things because we no longer have an overview of all the factors that affect that value. And then we speculate – or rather, gamble – on the change of those values to create even more "value." Madness. What is something worth, really?

It gets tricky when we value something that we cannot compare to something else. We only know what is good when we have a feeling for what is bad. We can only appreciate a time of peace by knowing what happens during war. This is the duality of existence: yin and yang. Value, therefore, also directly implies what we *don't* value.

In this way, the interpretation of what something is worth depends on your perspective, which is adjustable. Just hold your breath for two minutes and rediscover how valuable air is. We usually think air is worth nothing because it is everywhere in abundance and, best of all, free. Let's say that air has *general value*: from the human perspective, it is valuable to everyone, even if it is present in abundance. The same could be said for clean water, for example. In this way, we can identify three general values that will always represent the foundation of any economic system.

Value of energy

If we introduce free energy into society, energy suddenly becomes abundant compared to other things, such as material resources. It will, therefore, initially become a lot cheaper. But does it then actually lose value? No, because from the human perspective, energy is fundamental to our existence and the development of our society. Energy, therefore, has general value. Even though free energy technology ensures abundant energy, we should be incredibly grateful for it, just as we should be for clean air and water. With this attitude, we hopefully will not abuse or waste the abundance of energy.

Value of resources

Like energy, all of the Earth's matter – the raw materials and all the products we create out of them – has general value. Our bodies derive from the Earth's soil, water and air; we cannot live without them. We eat from the land and build our homes from rock and wood. To rediscover the Earth's generosity, consider that all resources currently in circulation are enough to meet everyone's basic needs. Especially if we cut our enormous waste and improve on recycling. With an abundance of energy plus the attitude that raw materials do not lose their value, we can recycle anything. Ultimately, every molecule, every atom we borrow from the Earth, regains the right to a new life. We can return this raw matter to nature or reuse it so we don't have to burden nature any further. This thinking is beautifully laid out in the book *Material Matters* by Thomas Rau and Sabine Oberhuber.[18]

In addition, free energy ensures that an enormous amount of raw materials can be saved from mining. Think of all the fossil fuels such as crude oil and coal for power generation, but also the toxic lithium for our batteries, iron ore for the steel of windmills, and so on.

Value of attention

Besides energy in its pure form and the Earth's resources, the third general value is our attention. What we give our attention to, grows. This is because our attention is also a form of energy that feeds our every activity. It is the attention we offer that counts – directly (as in healthcare) or indirectly (through a product or service).

The value – both in quality and quantity – of our attention is more challenging to express in numbers than, say, an amount of raw materials. It is perhaps the most subjective form of value because it involves human-to-human interaction. But as long as there is a (local) economy where value is expressed with a currency, we will value different occupations differently.

Free energy technology, as you can see, isn't necessarily going to change how we view human attention and different jobs. But free energy will make it easier to meet everyone's basic needs. That will at least allow us to revalue attention as one of the main pillars of the economy.

A new civilization

The following possible implications of free energy on society are a product of the three general values discussed above: energy in its pure form, Earth's resources and human attention. Of course, these implications do not equal predictions, but they paint a picture of a new civilization in which humanity flourishes on a new level. The possibilities are endless, so let your imagination run wild. Crucially, with every implication comes the consideration of balance with nature.

Production

Once clean energy is abundant, producing everything we need becomes much easier and cheaper. We have the energy to make practically anything we want anywhere, anytime. As noted earlier, we already have the raw materials in circulation (including products and waste) to make all the new things we want. New raw materials, such as the element carbon – for which we already have the most diverse applications, from building blocks to batteries – are becoming sustainable and easy to extract. And by making products truly durable again so that they last a lifetime, we will also need far fewer raw materials. Companies will no longer struggle to survive or make huge profits by flooding the market with cheap, low-quality products. This is because they have the energy to survive even with small batches. The path is clear for consumers to stop choosing mega-corporations that go only for big profits, usually accompanied by useless energy consumption, wasted resources and cheap labor. Products will again be of lasting quality, like the light bulb that has burned almost continuously for over a hundred years.[19]

Water, food and agriculture

The most critical factor for the availability of clean drinking water worldwide is (lack of) energy. However, desalination of seawater becomes very easy with free energy, and any coastal area may be supplied with enough water. Transporting the water inland is also feasible if enough energy is available. Moreover, water treatment plants can thoroughly clean water with abundant energy, requiring fewer or no chemicals. Even in very remote areas, if necessary, we could extract water from the air, provided we have enough energy.

Even the cultivation of just about any fruit and vegetable suddenly becomes drastically less capital-intensive. With free energy, we could grow tomatoes in a greenhouse anytime, anywhere – perhaps even in the desert. The latter is not necessarily wise, though. Tomatoes already grow very well in naturally occurring environmental conditions. We should let natural and organic cultivation take its course as much as possible. Because transportation with free energy is much cheaper and more sustainable, we could export tomatoes when they are in season like we already do, but without a "carbon footprint." This stimulates the local economy, making the world a little happier with a unique and good product. However, since ecology thrives best on what is produced locally by nature, the focus should be on local production to minimize unnecessary transportation.

As we become more aware of how nature thrives best in our environment, it creates a framework in which we can use free energy to its fullest potential. Our nutrition will be in tune with the local environment and season, precisely as nature intended. Biodynamic agriculture,[20] food forests[21] and permaculture[22] are examples of already existing techniques that we can use in collaboration with nature to make healthy food abundant for everyone.

Transport

With free energy, eventually, no vehicle will ever need refueling. This has enormous implications. Electric cars will never again have to stop to recharge. Cargo ships will either become wholly fuel-free or be replaced by airplanes. Airplanes will initially become electric and never have to refuel again. Then, new forms of electric propulsion will become available – such as the phenomenon of anti-gravity mentioned earlier – which will make the jet engine obsolete. Airplanes will become many times faster and quieter. Also consider the drastic reduction in environmental pollution due to eliminating exhaust fumes. Traveling to the other side of the world and transporting goods within a fraction of today's travel time will become cheap. Pause for a moment to imagine what the world will look like.

The world will get a lot smaller. People can suddenly go anywhere, including places they could never go before. Again, it is essential to strike a balance and ask ourselves with what intention we transport things and ourselves. After all, mass tourism is already a considerable problem in certain places, contributing to the pollution and destruction of nature. You can imagine that if everyone travels around all day, it can become extremely crowded in the air, on the road and at places of destination.

This means we must become more aware of other cultures, ways of life, and how nature thrives locally. In short, human action must be accompanied by a greater awareness of the environment and individual and collective goals of life. Yet, at the same time, we may respect and enjoy the ease of transport and no longer suffer a guilt complex about it.

Housing

Since the Industrial Revolution, there has been massive urbanization in more and more parts of the world. There are several reasons for this.

Initially, new industries in the city provided many jobs. Many people traded their relatively harsh rural existence for what they saw as an easier life in the city. Often, a steady job and a stable income were guaranteed. The population growth in the towns that accompanied this industrialization created much more demand for – and supply of – new goods and services, which in turn had a reinforcing effect on the economy and employment. With increasing urbanization, cities became a melting pot of cultures, ideas and opportunities. In the process, the city offered better education, healthcare, housing, recreational opportunities and a more prosperous social life than the villages in rural areas. Urbanization was a self-reinforcing effect in all aspects, so moving to the city became more attractive – especially for the young.

Soon, however, the adverse effects of urbanization started to reveal themselves. The cities became too crowded, so transportation and housing became problems. The initially attractive aspect of employment became saturated; unemployment went on the rise. Poor neighborhoods arose around the prosperous center, where healthcare and hygiene became issues and crime became a source of income. Overall, the pressure in the city increased, the infrastructure became too tight, and congestion and pollution became the norm.

The urge to achieve a better life in the city has been firmly rooted in human consciousness since the Industrial Revolution, despite the dire problems we started seeing. But with free energy, new opportunities to let society thrive can emerge anywhere, so people do not necessarily have to move to a city. Villages and communities can become fully self-sufficient, gaining the ability to take over any convenience and amenity that a city has – from agriculture and healthcare to education and recreation. As people distribute themselves more evenly among big cities, towns, villages and small communities, the pressure in the big cities decreases – resulting in a better balance between nature and the built environment.

Healthcare

Human attention heals. As we discussed earlier, the human body has fantastic self-healing abilities. As we become aware of our body's (subtle) energetic nature, we can prevent disease more effectively or tackle it at its source. We are beginning to realize that our body, like everything else in the universe, breathes life force. Thus, the body can be seen as free energy "technology" that we can consciously engage with our attention. People may primarily become their own healers, and healthcare no longer needs to be a billion-dollar industry.

The medical care that would be left would employ new technologies and therapies that work on the energetic rather than just the chemical level. But even before a diagnosis or treatment is prescribed, human attention remains the essential prerequisite for healing. Therefore, healthcare should never degrade into an assembly-line factory where people are received and treated by robots, a trend that has already started in many places, even though we may have abundant energy to do so. Healthcare, and care in general, implies genuine human attention.

Work and leisure

In the transition to a world where free energy is integrated, we will likely lose a good portion of the current job market. Many jobs will simply become obsolete. Consider, for example, the mega-corporations that now control our energy supply. Once free energy allows citizens to decide how much energy they generate and consume, we will no longer need an intermediary – except those distributing and maintaining the energy technology. As the economy decentralizes accordingly and ordinary citizens regain control of local finance, many jobs in today's financial sector and government will also be eliminated.

At the same time, free energy can create countless new work opportunities. Think of repairing the damage we have done to nature: cleaning up the oceans, restoring fertile soil, breaking down and recycling everything we no longer need – offshore wind farms, nuclear and coal power plants, fuel cars – and designing and building a society that is harmonious with nature. With an abundance of energy, it becomes a lot easier to make any meaningful ideal come alive, and therefore, there will certainly be no shortage of "work."

Again, the argument of free energy allowing us to meet our basic needs more cheaply and sustainably is key here. For many people, it means they don't have to earn as much money to make a living, and as a result, in addition to new work opportunities that free energy provides, we could be left with a surplus of free time. We call it "free time" because we don't have to devote this time to "work." However, this implies that we are not free at work – which is just one way of seeing things. In essence, you are always free. The question of what you will do with your "free time" may become mute. Instead, a new question may arise: What is the expression of a life in freedom?

Peace, freedom, responsibility

We have seen that free energy can do much good, but we cannot take the abundance of energy for granted. Free energy is not a license to further distance ourselves from each other and our environment, to do whatever we want with our expanded freedom. On the contrary, it requires us to be more conscious of our intentions and actions. We may become sovereign again, independent of a manipulative power structure. However, with great freedom comes great responsibility. Like yin and yang, the two aspects go hand in hand. If we want global freedom and peace, we must be conscious of the freedom and limits of all beings around us.

This chapter does not provide a ready-made answer as to what a new free energy society or economy looks like. A great deal will have to be reinvented or rediscovered. But hopefully, this book will inspire you to start working on what you find most meaningful and valuable. To discover what that is, it is vital to acknowledge the place from which you are looking – to truly know you – all the more reason to delve a little deeper into the 'phenomenon' of consciousness in the final chapter.

1. ourworldindata.org/water-access
2. ourworldindata.org/energy-access
3. ourworldindata.org/world-population-growth
4. ourworldindata.org/renewable-energy
5. ourworldindata.org/extreme-poverty
6. nos.nl/artikel/2153210-acht-rijksten-bezitten-evenveel-als-armste-helft-wereldbevolking
7. wir2022.wid.world/executive-summary/
8. society4th.org/product/society-4-0/
9. www.britannica.com/event/hyperinflation-in-the-Weimar-Republic
10. www.ecb.europa.eu/paym/digital_euro/html/index.en.html
11. www.investopedia.com/articles/investing/021816/why-governments-want-eliminate-cash.asp
12. www.youtube.com/watch?v=CiHXwWVEO3A
13. www.youtube.com/watch?v=7gCUh2ls0Yc
14. www.weforum.org/agenda/2022/09/my-carbon-an-approach-for-inclusive-and-sustainable-cities/
15. betalenmetflorijn.nl/
16. bofjoy.net/
17. charleseisenstein.org/essays/sacred-economics-money-the-gift-and-society-in-the-age-of-transition/
18. thomasrau.eu/en/material-matters/
19. interestingengineering.com/science/everlasting-lightbulbs-exist-ed
20. www.biodynamics.com/what-is-biodynamics
21. projectfoodforest.org/what-is-a-food-forest/
22. www.permaculturenews.org/what-is-permaculture/

Chapter 7: Consciousness is the key

"You can never solve a problem on the level on which it was created."

Albert Einstein

We have arrived at a time in history when many systems essential for humanity to thrive – finance, food, climate and energy – are in a state of strain like we haven't seen before. We are reaching the limits of these systems on a global scale. It is not that there are no solutions that could take us a step further in evolution, but they are withheld for more profound reasons. Einstein put it very aptly, as seen in the quote above. One could say that we are trying to solve problems from a perspective that caused the problems in the first place.

To solve a problem, one must approach it from a broader or 'higher' perspective. That is, at a higher level of consciousness. Until we realize that the source of our problem lies not outside us but somehow within ourselves, free energy will prove to be only a false solution. This would be the case when we use an abundance of energy for weapons more cataclysmic than the atomic bomb, for example. Or when we all suddenly get a fast car, a big house, and the latest stuff while giving nothing in return.

What can we do to pave the way to a world of energy abundance?

Self-realization

The picture of free energy we have outlined shows much resistance on the pathway to its introduction into society. If we look at the source of this resistance, it is often about power and fear. These traits are inherent in human experience. However, they seem to have taken over society's reins because we have not been conscious enough of them. The institutions in the world that make the introduction of free energy difficult or even impossible are the symptoms of an unconscious state of being in the human mind. The first place we encounter this unconscious state – and the best place to illuminate it – is within ourselves. The experiences of power and perpetration on the one hand and fear and victimhood on the other invite us to become aware of who and what we really are.

Who, then, are we really, and why are we here? Are we only a tiny lump of atoms accidentally capable of reflecting on its existence at the mercy of an indifferent and desolate universe? Or is there more than meets the eye, and is there a reason for our existence? We must begin to ask these essential questions again. The deeper we let them sink in, the more likely is the realization that what we call 'I' – I as a person, I as a man or woman, I with my background and past – is but a very narrow window through which infinite consciousness – what you really are – shines. In other words, when we become aware of our true essence, positive and negative experiences are just little waves on an immeasurably deep and peaceful ocean of being.

This is the funny thing about approaching the subject of consciousness. On the one hand, there seems to be a progression in life in which we become increasingly aware of our inner workings and their outer reflections. On this level – the level of you as a person – becoming conscious takes patience and courage. It involves facing your darker side and integrating it into your life to become whole as an individual. We could call this path "self-development," and there are endless courses, books, therapies and practices to explore it.

Paradoxically, however, the consciousness that you are is already fully aware and awake. This needs no development at all – only the realization that this is true. From the perspective of pure consciousness, life somehow unfolds spontaneously with remarkable peace and effortlessness, since consciousness itself is never tied to the outcome of the play of existence. Every so-called problem is allowed to be resolved, like watching a cloud evaporate in the blue sky. This is the essence of spirituality, the uncovering of our true nature – as consciousness – which is much grander than you can imagine.

What is abundance?

In the previous chapter, we looked at what abundant clean energy can do for the world on a physical level. We could call that the "physical manifestation of abundance." There, the main point was that there is enough on this planet for everyone. Enough energy, space, resources, attention, food, clean water and sustainable transportation for every living being.

At the same time, abundance is an experience for which free energy is not a prerequisite. It is vital to realize that the meaning and reality of abundance depend on how you view yourself and the world. Think of the things that you feel grateful for, and abundance will be real.

Explore for yourself what abundance means to you and how it feels. Where do you experience abundance in your life as it is now? And looking at the world outside, what might abundance mean to you in relation to your environment? What would you do if all the energy you ever needed was directly available? Take a moment to relax and visualize your idea of abundance. Become aware of the feelings this vision instills.

This thought experiment serves to open up a space in your mind for free energy to thrive in. After all, every new invention is born into this world – made physically manifest – from intuition and imagination. For those who feel inspired to bring free energy technology forward, the following paragraphs provide some guidance.

The practical side

In a practical sense, you can do much to expand the field of free energy. If awareness of your true nature comes first, then awareness of free energy in society comes second. Delve further into the research and increase your knowledge about it. Start a conversation with family, friends, and colleagues. Ask them what they would do with an abundance of clean energy. What does abundance mean to them? Tell them about your perspective, but don't try convincing anyone. Some people may not respond positively, and that's fine. Simply planting a seed by asking critical questions can have a significant effect. If you want to go bigger, inform policymakers on free energy. Involve the media. Use your network to spread the word, join a like-minded group, or start one yourself.

If you have an aptitude for engineering, delve into a free energy technology that you find fascinating. Remember that the internet is packed with nonsense, but you will find very useful and inspiring information among it. Start experimenting and connect with people doing the same. If you are interested in harvesting zero-point energy on an academic level, see the list of scientific articles in the bibliography at the end of this book.

As mentioned earlier, no ready-made and verified manuals exist yet for a free energy device. It is best to verify the working principle yourself to make it into a working prototype. It is, unfortunately, easy to get lost in the nitty-gritty of technical details, so zoom out occasionally to get the bigger picture back into view. Ask for help and funding – plenty of people are willing to help you – and be careful with any expectations to become the messiah who saves the world. This world doesn't really need saving. If it needs anything at all, it needs people living as their true self.

Conscious investment

One of the most considerable resistances in the world of free energy is money, as we have seen earlier. Vast profits are made from energy generation and trade, making it a major pillar of centralized power. But money also plays a crucial role at all lower levels of society when it comes to radical new inventions. For example, in developing a free energy device, we often see the problem that a researcher needs more financial resources to conclusively demonstrate that his device works but only receives financial support when he already can.

If you are an entrepreneur or investor; break this vicious cycle. Use your intuition and common sense to determine where your capital can flow best to support the free energy movement or to support a specific inventor, even if there is no officially verified prototype yet. Pay close attention to your intention. Is making a profit your primary goal, or do you want to save the world? Or are you simply following your greatest inspiration?

Provide a safe physical, mental and emotional environment where free energy research can thrive. Mutual trust, transparency and honesty are priority number one. It often turns out that, especially since we are dealing with major paradigm shifts, the gap between entrepreneurs and engineers can only be bridged by connecting from human to human. If you like bridging these gaps and are inspired by free energy, then you know what to do.

How do we market free energy?

We probably won't see a free energy device on sale at the hardware store anytime soon. The technology is too disruptive for that – it doesn't fit in the economy as it is. So, how does one market such a product?

We have already suggested in the previous chapter that a self-sufficient community with a relatively small circulation of free energy devices (and a local economy) can become independent of central energy distribution. Thus, several communities can gradually become energy-independent and serve as examples for the rest of society. That may be the smoothest and safest route (without a profit motive), but other options may exist.

Marketing a free energy device more openly in a conventional way could still work. However, this has not yet succeeded, often because of the investors' and inventors' widely diverging interests but also because of government and corporate intervention. Let's suppose everyone in your free energy company is on the same page. In that case, working "under the radar" may be possible to ensure that a working prototype is tested simultaneously in multiple locations. Then, the device can be safely given to the next group of people so the business can grow. After a while, the technology can be declared "free" and replicated as desired by individuals or other manufacturers.

Of course, producing a free energy device comes with a price, but the price must be manageable to ensure bottom-up adoption. Making exorbitant profits should never be the motive. If a free energy producer starts charging the same price per kWh as the traditional energy supplier, the consumer still pays a considerable portion of its money (labor) on energy. History teaches us that as soon as people get dollar signs in their eyes, for whatever reason, the introduction of free energy stagnates and fails. The level of consciousness of this old paradigm is incompatible with the new. The intention of introducing free energy must be for the world, not the company. Only then can the company serve as a vehicle to spread the technology effectively.

This brings us to another important point: what to do with intellectual property?

Patents versus open source

Usually, new inventions are patented as soon as possible to protect the inventor's intellectual property. Anyone else is prohibited from copying and bringing the invention to market. Patents used to be even more critical than today, as they have often represented a kind of seal of quality. Without a patent, any new invention was deemed unreliable and thus of little value.

Inventors of free energy technology ran into quite a bit of trouble in the past when applying for patents. Earlier, we saw that patent offices often refused to consider the applications of free energy pioneers. Their excuse was that there was no accepted physical theory to explain the invention. If the applications were accepted, they were not infrequently labeled by governments as dangerous to (inter)national security and subsequently declared secret.

Many inventors, partly for this reason but sometimes also out of self-interest or the interests of investors, withheld crucial details in the patent. As a result, the technology was hardly reproducible. All in all, patents have caused much resistance and stagnation.

For these reasons, open sourcing the invention might be a better idea. In open source, at some point in the development of the technology, all technical specifications are released on the internet so that the technology can be replicated and improved by anyone. Once the information about a working free energy technology is clearly formulated and widely distributed on the Internet, the inventor and his company can no longer be the target of sabotage. The technology has been birthed into the world and has become unstoppable. This, of course, is the true spirit behind the term free energy. The energy ultimately belongs to no one, and no one can acquire a monopoly position with it, while everyone can benefit from it.

Energy abundance

What can we conclude? We are on the brink of the most profound transition in human evolution to date. Much is at stake, or rather, everything is at stake. We either continue the current path of so-called "sustainability," with all its disastrous implications – seas full of windmills, lands full of solar panels, continued pollution of nature, continued dependency on nuclear and gas power plants – or we open ourselves to a world of actual abundance. That is not to be trifled with, as we have collectively lived a mindset of scarcity and struggling to survive for thousands of years. In this time of remarkable confluence of crises lies the greatest challenge and, therefore, an incredible potential for growth.

The transition to a world of abundance is partly a matter of technology. Free energy – tapping into an inexhaustible source of energy available everywhere and at all times – is undoubtedly the invention of the millennium. With it, every conceivable system and every aspect of our existence will be elevated to an unprecedented level. If this technology exists – and if you have come this far, you can judge for yourself – all that remains is the question: Are we ready for it?

That question and its answer are less technical, leaving our fingers pointing inward. The changes that free energy (or lack thereof) will bring about are so profound that we are thrown back to the essential questions of our existence. Therefore, the extraordinary evolution man is about to make is not an evolution of technology but of consciousness. And out of higher consciousness naturally arises thinking and acting that aligns with nature's inherent harmony, a climate in which free energy – energy abundance – can flourish.

Acknowledgments

This book is part of an exciting journey that I get to take with many great people. So here, I would like to put them in the spotlight.

First, I thank my friend Casper Boom, with whom I founded Energy Abundance. He was like a brother who shared the burning fire I felt for free energy. Working together taught me a lot about myself – sometimes it was like looking in the mirror – and we created many adventures to raise awareness of free energy.

Secondly, I cannot thank Coen Vermeeren enough. Without his mentorship and inspiration, the idea of writing a book might never have occurred to me. He took the book to the next level with incredible attention to detail, from content to editing, which has proven invaluable. Thank you.

The one who has donated much of her time and love to the book in between, with whom I can philosophize to the depths and who has always supported me, is my mother. Moms, without your help, this book would be a rough stone, both in terms of content and language – much love to you.

I also thank all the people who have made a financial contribution to Energy Abundance, small or large. If you are one of these contributors and reading this, please know that your support has been vital for me to research and increase awareness of free energy. Thank you very much.

Many people have also helped and motivated me by sparring together about free energy and its grand potential. Some have also contributed directly to this book by providing corrections and tips. Thank you, Susan Manewich, Jeane Manning, Ivo Bonants, Koen van Vlaenderen, André Norel, Frank Bonte, Philippe and Romein Metz, Jan-Willem van den Boogert and dear Arabella Meyer. Thanks also to Peter from Team Obelisk for his patience (with me) in designing the cover.

There may be people not mentioned here that I should have: please know that your help and love are greatly appreciated.

This book is also dedicated to all the pioneers in the field of free energy. Your hard work has lit a Promethean fire. You may not have experienced the era of abundant clean energy on Earth yourself, but your inspiration lives on and I am sure your shared vision will once be reality.

Finally, I thank you, the reader, for reading this book. I hope to have inspired you in any way to embrace a world of abundance.

Glossary

Alternating current (AC)

An electric current that goes "back and forth" in an electrical circuit. With alternating current, the voltage (like the amperage) goes up and down over time in a wave motion.

Capacitor

An electronic component consisting of two electrical conductors with an insulating material in between. A capacitor can store electrical charge, by which it can convert an electric current into an electric field and vice versa.

Charge (electric)

The property of a particle to be either "positive" or "negative" electrically. For example, an electron has one negative charge, and a proton has one positive charge. Positive and negative charges attract each other, and equal charges repel each other.

Coefficient of Performance (COP).

A number that represents the efficiency of a system. It is the ratio of the amount of useful energy a system provides to its energy consumption. For example, a COP of 1 means 100% efficient, and a COP above one represents over-unity. *NB: COPs above 1 are typical in heat pumps and refrigerators, but these devices do not generate free energy.*

Coil (electronic)

An electronic component consisting of a wire wound in a spiral shape. A coil can convert an electric current into a magnetic field and vice versa.

Current

The amount of electrical charge flowing through a conductor per unit of time, measured in Ampère.

Direct current (DC)

An electric current that flows in only one direction through an electrical circuit. With direct current, the voltage in a circuit is usually constant.

Electric field

A field in space associated with a difference in electric charge between two points in space. An electric field has a direction from positive to negative charge.

Electromagnet

An electronic component that consists of a coil wrapped around a metal core (often containing iron). A magnetic field is created when a current flows through the coil, causing the metal core to become magnetized. Turning the current on and off causes the magnet to turn on and off.

Electromagnetic field

A field in space in which an electric and magnetic field are combined. All forms of radiation (e.g., visible light or radio waves) are vibrations in the electromagnetic field.

Impulse (electric)

A short, powerful electric shock wave in an electrical circuit or electromagnetic field. An impulse can be measured as a voltage or current spike.

Magnetic field

A field in space associated with the rotation (spin) or motion of electric charge. A magnetic field from rotating charges has a direction from north (clockwise spinning charge) to south (counterclockwise spinning charge).

Over-unity

An efficiency of a system above 100%. The system provides more (useful) energy than the system consumes to operate. This is possible when it is an open system tapping into an abundant energy source such as the ether.

Vacuum tube

An electronic component consisting of a glass tube from which air has been sucked out, containing a positive pole (anode) and a negative pole (cathode). By heating the cathode, electrons can "jump" from the cathode to the anode, but not vice versa. This made the vacuum tube the first electronic rectifier (diode). The vacuum tube was widely used in early radio equipment and is also called an electron tube or radio tube.

Voltage/Volt

A unit of measurement representing electrical tension or pressure, measured in Volts. Electrical tension means a difference in electrical charge between two points, giving rise to an electric field.

Zero-point energy (field)

The scientifically common term for the vast sea of background energy of the universe – the energy source that free energy technologies can tap into. Other names include ether/aether, vacuum energy, (quantum) vacuum fluctuations, quantum foam, Sea of Dirac, quintessence and cosmological constant.

Further reading

Books

- Manning, Jeane & Garbon, Joel. *Breakthrough Power: How Quantum-leap New Energy Inventions Can Transform Our World.* USA: Amber Bridge Books, 2011.
- Manning, Jeane & Manewich, Susan. *Hidden Energy: Tesla-Inspired Inventors and a Mindful Path to Energy Abundance.* USA: self-published, 2021.
- Oswalt, Ellis. *Tesla's Words.* UK: self-published, 2020.
- Vassilatos, Gerry. *Secrets of Cold War Technology.* Eureka, CA: Borderland Sciences, 1996.
- Zohuri, Bahman. *Scalar Wave Driven Energy Applications.* New York: Springer, 2019.

Patents

- Patrick, S. L. & Bearden, T. et al. (2002). *Motionless Electromagnetic Generator* (U.S. Patent No. 6,362,718).
- Kapanadze, T. (2008). *Independent Energy Device* (Int. Patent No. WO2008/103129).
- Tesla, N. (1901). *Apparatus for the utilization of Radiant Energy* (U.S. Patent No. 685,957).
- Tesla, N. (1901). *Apparatus for Utilizing Effects Transmitted through Natural Media* (U.S. Patent No. 685,956).
- Tesla, N. (1901). *Method of utilizing Radiant Energy* (U.S. Patent No. 685,958).

Articles

- Nikola Tesla, *"The Problem of Increasing Human Energy"*, Century Magazine, 1900.
- Peter Lindemann, *"The World of Free Energy"*, 2001. free-energy.ws/lindemann-1.html

Scientific articles

- Moddel G, Weerakkody A, Doroski D, Bartusiak D. *Optical-Cavity-Induced Current. Symmetry*. 2021; 13(3):517. doi.org/10.3390/sym13030517
- Santhanam P, Gray DJ Jr, Ram RJ. *Thermoelectrically pumped light-emitting diodes operating above unity efficiency*. Phys Rev Lett. 2012; 108(9):097403. doi: 10.1103/PhysRevLett.
- King MB. *Water Electrolyzers and the Zero-Point Energy*. Physics Procedia 20. 2011; 435-445. doi.org/10.1016/j.phpro.2011.08.038.
- van Vlaenderen K. *General Classical Electrodynamics*. Universal Journal of Physics and Application. 2016; 10(4): 128-140.
- Macken J. *A quantum vacuum model unites an electron's gravitational and electromagnetic forces*. 2021; 10.13140/RG.2.2.20007.68003.
- Takagi O, Sakamoto M, Yoichi H, Kawano K, Yamamoto M. *Scientific Elucidation of Pyramid Power: I*. Journal of ISLIS. 2020; 38(2): 130-145.

Websites

News, information and awareness:

aetherometry.com
breakthroughenergyproject.org
globalbem.com
hyiq.org
lenr-canr.org
newenergymovement.org
newenergytimes.com
nulpuntenergie.net
padrak.com/ine/
rexresearch.com/1index.htm
teslauniverse.com

theorionproject.org/en/research.html
zpenergy.com

Forums and (active) research:

am-innovations.com
energeticforum.com
integrity-research.org
jnaudin.free.fr
overunity.com
overunityresearch.com
pacenet.homestead.com

Image credits

Every image in this book is either the work of the author or selected from the public domain, except for the following:

- Figure 2.2, left: © OneTesla, cropped.
  commons.wikimedia.org/wiki/File:OneTeslaTS_DRSSTC_Tesla_C
  oil_closeup.jpg
  License: Creative Commons Attribution Share-Alike Int.
  creativecommons.org/licenses/by-sa/4.0/legalcode
- Figure 3.1: © Emok.
  commons.wikimedia.org/wiki/File:Casimir_plates.svg
  License: Creative Commons Attribution Share-Alike Unported
  creativecommons.org/licenses/by-sa/3.0/legalcode
- About the author, first image: © Juul Geluk Fotografie
  www.juulgeluk.com/
- Pag 46: Britannica.com. Publicity photo of Nikola Tesla in his
  laboratory in Colorado Springs, Colorado, in December 1899.
  Tesla posed with his "magnifying transmitter," which was
  capable of producing millions of volts of electricity. The
  discharge shown is 6.7 metres (22 feet) in length.

About the author

Curiosity is what drives Karsten van Asdonk. From a young age, he has had many passions, from human anatomy to space exploration, which he explored to the core. He studied biomedical engineering and applied physics at Eindhoven University of Technology, where he worked toward a specialization in nanotechnology. His primary interest, however, had grown to be outside his studies, in what current science cannot yet explain. This includes topics such as the interconnectedness of consciousness and matter, quantum physics, spirituality, and free energy. He aims to raise awareness on the cutting edge of science and spirituality, where these two fields already meet.

The logo of Energy Abundance, lovingly designed by Isabel Camps, represents the toroidal energy field with the heart at its center.

Website: www.energyabundance.nl
E-mail: info@energyabundance.nl

ENERGY ABUNDANCE